GOVERNING THE ATOM

ENERGY AND ENVIRONMENTAL POLICY Series:

Technology and Energy Choice, Volume 1
John Byrne and Daniel Rich

Energy and Cities, Volume 2
John Byrne and Daniel Rich

Politics of Energy R & D, Volume 3
John Byrne and Daniel Rich

Planning for Changing Energy Conditions, Volume 4
John Byrne and Daniel Rich

Energy, Land and Public Policy, Volume 5
J. Barry Cullingworth

Energy and Environment:
The Policy Challenge, Volume 6
John Byrne and Daniel Rich

Governing the Atom:
The Politics of Risk, Volume 7
John Byrne and Steven M. Hoffman

Edited by
JOHN BYRNE &
STEVEN M. HOFFMAN

GOVERNING THE ATOM

The Politics of Risk

ENERGY and ENVIRONMENTAL POLICY

VOLUME 7

Routledge
Taylor & Francis Group

LONDON AND NEW YORK

First published 1996 by Transaction Publishers

Published 2019 by Routledge
2 Park Square, Milton Park, Abingdon, Oxon OX14 4RN
52 Vanderbilt Avenue, New York, NY 10017

Routledge is an imprint of the Taylor & Francis Group, an informa business

Library of Congress Catalog Number: 96-12413

Library of Congress Cataloging-in-Publication Data

Governing the atom : the politics of risk / edited by John Byrne, Steven M. Hoffman.
    p. cm. — (Energy and environmental policy ; v. 7)
    Includes bibliographical references and index.
    ISBN 1-56000-834-2 (alk. paper)
    1. Nuclear industry—Environmental aspects. 2. Nuclear industry—Safety measures. 3. Nuclear energy—Government policy. 4. Nuclear power plants—Accidents. I. Byrne, John, 1949– . II. Hoffman, Steven M. III. Series.
HD9698.A2G628   1996
333.792'4—dc20   96-12413
                                          CIP

ISSN: 0882-3537
ISBN 13: 978-1-56000-834-7 (pbk)
ISBN 13: 978-1-138-52452-1 (hbk)

Contents

v

Introduction

John Byrne and Steven M. Hoffman

In his 1961 farewell address, then U.S. President Dwight D. Eisenhower informed the nation that its successful development of an atomic weapon was made possible by a common effort of the military, science, and corporate sectors. The integration of civilian and military, scientific and industrial knowledge, organization and interest had led to the forging of what he called a "military-industrial complex." This complex, he observed, had become the backbone of a new era in which human hopes and dreams could be met by institutional cooperation and maximum use of scientific expertise. But he also warned that this complex could test and even corrode the democratic aspirations of society. And he noted that the technological triumphs possible from institutional cooperation could yield highly dangerous, as well as positive, outcomes. It was the challenge of our era to see that democracy prevailed and the activities of the military-industrial complex[1] served rather than threatened our future. Eisenhower's initiation of the Atoms for Peace Program earlier in 1953 was an expression of his optimism that the challenge could be met.

The widespread use of nuclear power to meet energy needs confirms that Eisenhower's optimism has been embraced by many

---

[1] While Eisenhower referred in his speech to a "military-industrial complex," he clearly recognized science as a third partner, emphasizing that we needed to be equally wary that "public policy could itself become the captive of a scientific-technological elite" (quoted in Kevles, 1987: 393).

1

in the new era. Unfortunately, his warning of threats to democratic society posed by the advance of a military-industrial-science complex has been less successful, gaining the attention of society only infrequently. This is especially true in the case of nuclear power whose development was borne out of and remains dependent upon the consortium of interests identified by Eisenhower. Despite overwhelming evidence of the corrosive effect that nuclear power development has had on democratic ideals and practice — see especially the chapters by James Jasper, Phillip A. Greenberg and Michael T. Hatch in this volume; despite a record of disregard for safety in order to achieve the technical possibilities of nuclear power — see especially the chapters by Cate Gilles, Phillip A. Greenberg, Carolyn Raffensperger and David Marples; and despite a highly authoritarian political economy and science apparatus that has grown up in support of the technology's diffusion throughout the world — see especially the chapters by Cecilia Martinez and John Byrne and Jong-dall Kim and John Byrne, as well as our own chapter; in sum, despite a history replete with danger signals, societies around the world have failed to subject the Nuclear Project to the public scrutiny warranted, arguably, for the most dangerous technology invented in our time. Regrettably, our era has preferred the message of promise over warning in the development of nuclear power.

True, this preference has only managed to prevail with *serious* reservations expressed about the enterprise. For instance, Three Mile Island and Chernobyl are symbols recognized throughout the world of how things can go terribly wrong in the case of this technology. And the spread of nuclear power has been accompanied in nearly all countries that rely on it by the rise of antinuclear movements that have, often effectively, stymied expansion of its use. But it would be a mistake to interpret these facts as evidence that this technology's advance has been checked, or that it is now satisfactorily governed by society, rather than dictating to it.

On the status of the Nuclear Project, there are now 424 reactors operating throughout the world.[2] The total capacity of these plants (337,518 megawatts — MWe) represents 6% percent of the world's electrical supply. While a relatively modest share of current total capacity, nuclear power's role in world electricity supply has grown substantially since 1970, when nuclear's share was barely 0.5 percent. A number of countries now rely on nuclear power for over 40% of their electrical needs, including Lithuania (87%), France (78%), Belgium (59%), the Slovak Republic (57%), Hungary (43%), Slovenia (42%), and Sweden (42%) (EIA, 1994). But this number will increase quickly too. Eastern Europe, Russia, Latin America, and Asia, in particular, have aggressive nuclear programs and ambitions. Of the 66 reactors (54,022 MWe) presently under construction, 23 will come on line in Asia, 17 in Eastern Europe, 4 in Russia, and 4 in Latin America.[3] Most are scheduled for operation before the end of the century. Both the existing and new reactors have average capacities that are twice or more the size of standard non-nuclear plants. Two countries' nuclear goals for the early 21st century will result in their reliance on this technology for 50 percent or more of national electrical supply. Japan, which currently relies upon nuclear power for 31% of its electricity, is projected to increase its nuclear capacity by 12.4 gigawatts (GWe) — to a total of 55.2 GWe in 2010. South Korea, currently the operator of nine units, projects a doubling of capacity, to between 13.0 and 16.1 gigawatts by 2010 (EIA, 1995: 52).

With the average price tag of a 1,000 MWe nuclear reactor at nearly $3.0 billion, the replacement cost for the existing nuclear system is approximately $1.0 trillion. With decommissioning costs included (see Raffensperger in this volume), the "cradle-to-grave"

---

[2]   Currently, only the continents of Antarctica and Australia have no nuclear plants.

[3]   The remaining plants under construction are distributed as follows: 10 reactors in Western Europe, 2 in Cuba and 6 in the United States.

investment in the Nuclear Project (in 1995 replacement dollars), is approximately $1.5 trillion. This is nearly twice the amount spent between 1970 and 1993 on all forms of development assistance throughout the world (World Bank, 1995). The 66 new plants will cost (with decommissioning included) roughly $240 billion, or more than the projected expenditures on development assistance from 1994 through the year 2000 (using actual expenditures for 1990-93 — see World Bank, 1995). In brief, worldwide commitment to nuclear power is and will remain substantial.

The continuing advance of the Peaceful Atom, notwithstanding the less than peaceful possibilities it engenders (as the recent negotiations over North Korea's nuclear ambitions illustrates), is testament to the power of nuclear energy's message of optimism to override its realities. But the darker side of the Project, identified in Eisenhower's warning, continues to be equally evident. Nuclear power development has been and is now determined by elite decisions in every country in which it is being pursued. There is no reason to believe that this will change in the foreseeable future. Indeed, as argued by several authors in this volume, it may not be possible for this condition to change. One issue is that, in most cases, the nation's military is a major (and frequently dominant) partner, underscoring the difficulty of imprinting a democratic structure of governance on the Project. The technology's inherently dangerous nature is a second problem. It means that society may have little choice but to rely on an apparatus imbued with secrecy, security considerations, and technocratic values to protect it against the catastrophic implications of "accidents" during its use. But the continued antidemocratic tendency of the Project's development also stems from what are likely to be the next-generation improvements in the technology. The era of the fast-breeder reactor is near and, with it, the prospect of a closed fuel cycle. From a scientific or technological point of view, this may be regarded as a triumph that advanced societies should endorse. However, for those concerned with democratic governance, this can hardly be taken as an unalloyed success.

In sum, the Nuclear Project is and will grow in size and importance. Technological optimism will be a propulsive force in its growth, as will its political, economic and cultural symmetry with the logic of technological society. The intent of this volume is to examine critically the social implications and demands that are associated with this technology's development and to identify and evaluate the type of society it fosters.

The first part of the volume examines the social structure required for the development and maintenance of the technology. Three chapters take up this theme.

The first chapter (Byrne and Hoffman) argues that the widespread adoption of nuclear power is predicated upon an *ideology of progress* which tends to ignore risks associated with complex technology in favor of the potential benefits presumed to arrive in its future. Disputing claims that the spread of nuclear power derives from either its economic or environmental advantages, we suggest instead that its desirability is based upon the conformity of the technology's ideology with that of technological society.

In Chapter 2, James Jasper documents and analyzes the policy regime that has underlain the state's promotion of nuclear power. Through a comparison of the nuclear histories of France and the United States, he is able to demonstrate the central role of policy and political institutions in the development of nuclear energy. In both countries, he argues that policy makers were caught up in the thrall of development, so much so that they engaged in a kind of "hortatory" policy strategy. That is, "planners and energy experts were thrilled by their predictions, excited at the possibilities, and so exaggerated them" in order to realize the possibility of nuclear power.

Another critical factor in nuclear power's appeal has been the power its designers and manufacturers have enjoyed, especially in the latter part of the 20th century. In the third chapter of the volume, Cecilia Martinez and John Byrne offer an analysis of the scientific and corporate networks which emerged in support of the technology. The authors point to the successful collaboration of science, industry and the nation state in nuclear power development as a foundation and model for postindustrial society and its emphasis on high technology-based growth. They argue that because this consortium of interests is a core social requirement for modernism, technological societies find it exceedingly difficult to impose democratic controls on the Nuclear Project.

The second part of the volume considers the risks, costs and dangers that societies have endured as part of what Alvin Weinberg has referred to as a "faustian bargain" (1972) — the availability of a limitless energy source in return for society's acceptance of the special political, economic and ideological demands of nuclear technology.

Cate Gilles' examination of Native American communities in the American Southwest (Chapter 4), the site of massive uranium mining and milling activities over the last forty years, is a stark reminder that the costs of nuclear power are oftentimes borne by those who were and are excluded from the corridors of power. The legacy of "nuclear colonialism" has, according to Gilles, left these communities with grossly exaggerated levels of cancers and other diseases, fouled their water supplies, and so degraded their lands that living on them can threaten life itself.

In Chapter 5, Phillip A. Greenberg offers a comprehensive examination of the range of societal risks posed by nuclear power. While a number of authors have examined the issue of nuclear risk, Greenberg places it in the much needed, but until now largely neglected, context of democratic governance. Specifically, he raises the question of who is to decide whether the risks of catastrophic accidents are acceptable? Challenging the view that public fears of nuclear power are irrational and borne out of ignorance, Greenberg offers an insightful analysis of why the public may be right in distrusting the technology, its advocates and its designated regulators. At stake in the debate, as he points out, is whether the public can retain to itself the right to judge the technology and define its risk, or whether experts will continue to successfully claim the right to decide for them.

The volume's survey of social costs and risks associated with nuclear power is completed with Carolyn Raffensperger's examination of decommissioning and decontamination (Chapter 6). She describes in detail the U.S. legal framework to safeguard the public and the natural environment from the inevitable risk of accident-free, nuclear plant operations — a waste stream of long-lived, radioactive waste composed of the structures and surrounding soil of each retired facility. In the U.S. case, this equals approximately 15 million cubic feet and will cost in excess of $130 billion (based on the decommissioning experience at the Shippingport reactor). Raffensperger reminds us that, in many respects, the battles being waged at Prairie Island, Minnesota and Yucca Mountain, Nevada over the storage of radioactive wastes are only opening skirmishes in what will be a long-term conflict for any society that joins the Nuclear Project.

The final section of the volume provides a review of nuclear power development in industrial and developing countries. With this review, the globalization of the technology and its implications can be considered.

Michael T. Hatch begins this review (Chapter 7) with a comparative analysis of political structures found in the United States, France and Germany charged with overseeing these countries' nuclear power programs. As he notes, these countries are and will be "the most influential in determining the . . . course of nuclear power in the West." Hatch argues that corporatist forms of political organization have accompanied these countries' efforts to develop nuclear power. More democratic and pluralist political impulses in all three countries are shown to have clashed repeatedly with the governing structures of this technology. He concludes that corporatist arrangements have lost efficacy in all three countries and pluralist demands for a more open political process to oversee nuclear power are perhaps unavoidable in the West.

While Western interest in nuclear power has waned, East European and Russian interest has not. David Marples' examination of the status and direction of nuclear power development in Russia and the Commonwealth of Independent States (Chapter 8) emphasizes the staying power of the technology, even in the face of deteriorating safety standards and poor operating conditions. While moribund for a period after the Chernobyl disaster, Marples finds that nuclear power is experiencing a resurgence in this part of the world. The higher leukemias, thyroid cancers and morbidity rates in Belarus and the Ukraine in the aftermath of the worst nuclear plant accident to date have not been sufficient to derail further use of this technology.

The volume concludes with Jong-dall Kim and John Byrne's assessment of nuclear power aspirations in East Asia (Chapter 9). Taking up the theme of the first chapter of the volume, they describe how energy planners in Japan, North and South Korea, China and Taiwan have been swayed by the promise of nuclear power, particularly its association with social progress. Their chapter indicates that the "nuclearization" of East Asia's energy policy has proceeded on a centralist political and economic basis, notwithstanding the significant diversity of political and economic systems represented by these countries. In this fast-growing market for nuclear energy (it currently accounts for nearly one-third of all new plant orders), the power of the technology to implant its own ideological orientation alerts us that the Nuclear Project is far from being completed.

References

EIA (Energy Information Administration). 1995. *International Energy Outlook.* Washington, D.C.: EIA.
_____. 1994. *World Nuclear Outlook.* Washington, D.C.: EIA.

Kevles, Daniel J. 1987. *The Physicists: The History of a Scientific Community in Modern America.* Cambridge, MA: Harvard University Press.

Weinberg, Alvin. 1972. "Social Institutions and Nuclear Energy." *Science*, Vol. 177: 27-34.

World Bank. 1995. *World Development Report 1995.* Oxford, UK: Oxford University Press.

PART I

The Social Structure of Nuclear Power

Chapter 1

The Ideology of Progress and the Globalization of Nuclear Power

John Byrne and Steven M. Hoffman

Introduction

As an energy source, nuclear power was not technologically feasible nor economically viable when it was embraced by the U.S. in 1946. It did not originate as an invention of enterprise, nor was there a market for its supply. In fact, the U.S. committed itself to the development of the "peaceful atom" 11 years before it would be successfully demonstrated. The national government sought to discover the advantages of the technology and to discount its costs in the *absence* of knowledge of its economic or technical practicality. When Lewis Strauss, a former chairman of the U.S. Atomic Energy Commission (AEC), announced that nuclear power would bring forward an energy supply "too cheap to meter," he signaled that, for this technology, social desirability would be decided in advance of performance, since his declaration of nuclear energy's economicalness was 17 years before the opening of the first commercial reactor (Byrne and Rich, 1986).

Despite the catastrophic accident at Chernobyl in 1986, the near meltdown at Three Mile Island in 1979, over 200 "precursors" to core meltdown accidents in the brief period of the technology's commercial use (Adato et al, 1987), and an industrial history worldwide of massive cost overruns, nuclear power continues to be evaluated in the "future tense,"[1] that is, in terms of what it will bring rather than what it has already wrought or what it requires from society to maintain operation. While enthusiasm is expressed more modestly today than in the heyday of its early promotion, support for nuclear power remains strong in several quarters despite its authoritarian politics, its failed economics and its dubious performance history. Thus, Mr. Ryo Ikegame, executive vice president of Tokyo Electric Power Company, one of the largest electric utilities in the world, recently offered this assessment of nuclear power in the only country that has suffered a nuclear attack (Taylor, July 1992: 32):

> [I]t rained after Chernobyl, and now it's cloudy —
> but we can see the sunny part of the sky. I'm
> rather optimistic about the future of nuclear power
> plants, because Japan has no oil, no coal, no gas —
> so we have to depend on nuclear, and this is good
> for the environment.

Nuclear development plans for Japan reflect this belief: over the next twenty years, Japan intends to add 38 more nuclear plants to its existing stock of 49 (for a total nuclear capacity of 40 GWe); 5 of these plants are already under construction and will begin operation by 1997 (*Nuclear News*, August 1992: 60-61; March, 1995: 32-33; June, 1995: 40). Japan is not alone in its commitment to nuclear power: as of December 1994, 66 nuclear plants are under construction or on order in 19 countries, the majority of which are scheduled for completion by the year 2001 (*Nuclear News*, March, 1995: 27-42).

---

[1]   This phrase is borrowed from David Noble, 1983.

Below we examine the continuing worldwide momentum for nuclear power development. It is argued that support for nuclear power is embedded in first, the modernist ideology of progress that equates economic growth and technological power with social success; and second, the "nuclear consortium" — comprised of the state, military, science and industrial apparatuses — which must be integrated in order to develop nuclear technology (Camilleri, 1984; Byrne, Hoffman and Martinez, 1989). Together the modernist ideology of progress and the nuclear consortium are argued to constitute a political economy of "technological authoritarianism" (Byrne and Hoffman, 1988). This political economy has been institutionalized in the core industrial countries and is now being "transferred" to the periphery and semi-periphery countries of the Third World.

Nuclear Power and the Industrial Idea of Progress:
The Case of the U.S.

Since industrialization, Western ideas of progress have equated social success with national wealth and scientific and technological prowess (Mumford, 1934). In this equation, energy has had a central role. Indeed, a routine assumption of industrial societies throughout the 20th century has been that higher energy consumption directly corresponds with higher orders of civilizations. As Aldous Huxley remarked, "because we use a hundred and ten times as much coal as our ancestors, we believe ourselves a hundred and ten times better intellectually, morally and spiritually" (in Basalla, 1980: 40). Basalla (1980) coined this idea of progress "the energy-civilization equation."

Not only more energy but more sophisticated technologies to produce, distribute and use energy are prized in the energy-civilization equation. In this vein, electric power plants have held special status as the highest stage of development yet achieved by Western civilization to realize the goal of cheap and abundant energy. Power plants are the ideal "abundant energy machines"

(Byrne and Rich, 1986) and in the energy-civilization hierarchy, nuclear power has been touted by many as the most advanced of our energy machines. William Laurence, an early American nuclear propagandist, captured the modernist attraction to the technology when he characterized atomic energy as a "veritable Prometheus bringing to man a new form of Olympic fire" (1940: 12-13) that would deliver "wealth and leisure and spiritual satisfaction in such abundance as to eliminate forever any reason for one nation to covet the wealth of another" (1959: 240). Weinberg echoed this sentiment when he branded nuclear power "a marvelous new kind of fire" (1972: 28) capable of providing "the solution to one of mankind's profoundest shortages" (1956: 299).

Virtually all industrial countries have actively pursued nuclear power programs in the latter half of the 20th century in the hope of securing a bountiful energy future. The U.S. has the longest running commitment to the nuclear dream and has built more domestic nuclear capacity and exported more nuclear plants than any other country in the world. Its nearly 50 years of experience with the technology, however, has hardly been problem-free. American nuclear "troubles" include a long list of accidents, unresolved waste disposal problems, the irradiation of Native American communities and lands, plant workers and neighborhoods adjacent to nuclear facilities, and never-met promises of low-cost energy supply (Carter, 1987; Byrne, Hoffman and Martinez, 1992; Gilles, this volume).

Perhaps the most telling evidence of nuclear's appeal to U.S. science and business elites is the enduring commitment to build and operate a nuclear complex in the country's Pacific Northwest. The history of the Washington Public Power Supply System illustrates the staying power of the nuclear idea of progress in the face of overwhelming evidence of failure. During the mid-1950s, the Pacific Northwest was presented with a "future-tense" crisis: the possibility of restrictions on the economic growth of the region conditioned by a lack of energy resources. Home to some of the world's largest hydroelectric dams operated by an integrated system of region-wide management, the area nonetheless enthusiastically embraced nuclear power as the proposed solution to this dubious dilemma. The region's nuclear dream was articulated in the so-called Hydro-Thermal Power Plan (HTPP) adopted in 1968 by a consortium of 108 investor-owned and public utilities located in six states in the Pacific Northwest. The HTPP proposed to add 41,400 megawatts of hydro- and thermal power — including 20 new nuclear power plants. It was a spectacular vision of energy abundance justified in appropriately fervid terms: "Increased use of electricity has contributed importantly to the emancipation of peoples from poverty and drudgery and to expansion in human capacity to live the good life . . . [T]o expand these benefits to a larger segment of our society will require more electricity" (quoted in Olson, 1982: 19).

A few years later the technology's advocates were forced to temper their dreams but by the early 1970s the Washington Public Power Supply System (WPPSS) nonetheless found itself trying to manage the simultaneous construction of five nuclear plants. As so often has been the case with nuclear projects, things did not proceed according to plan, even though premier engineering and financial companies managed WPPSS. For the project, Bechtel and the "Big Four" reactor vendors — General Electric, Westinghouse, Babock and Wilcox and Combustion Engineering — were responsible for the design and construction of the five plants, while Merrill Lynch, Paine Webber, Solomon Brothers and Blythe Eastman Dillon managed the financial transactions (Byrne and Hoffman, 1992). Despite this impressive gathering of engineering and financial acumen, on January 22, 1982, construction was terminated on two of the five plants and by 1983, two other plants had been mothballed. Only one plant, with a rated capacity of 1,154 MWe was ever completed. All told, the System defaulted on municipal bonds representing $6.7 billion in principal and some $23.8 billion in interest. This represents, by the far, the largest default in the history of the American municipal bond market. The System collected over 30 years of construction delays while completing only one plant capable of delivering continuous electric power to the region. According to estimates made in the late 1980s, the WPPSS plants have been largely responsible for the region's 700 percent increase in wholesale power rates and 250 percent increase in household electric rates (Comptroller General, 1984: 7; Northwest Power Planning Council, 1988: 2).

Despite this record, the region's energy planners persisted in their favorable treatment of nuclear power. The 1986 Northwest Conservation and Electric Power Plan, for instance, argued that WPPSS Nuclear Plants 1 and 3 could still be completed, providing the region with some $630 million worth of net benefits; these presumed social benefits, were, of course, contingent upon the commitment of a $2.8 billion investment to complete the plants. Only very recently have these same planners accepted the fact that nuclear power will not constitute a major source of new energy for the region. Thus, on May 13, 1994, the System's managers voted to terminate WNP-1 and 3 effective January 13, 1995. Yet, even at this stage in a long history of failure, some in the System are trying to salvage the technology. In a proposal that underscores the arbitrariness of the distinction between military and civilian uses of the atom, the WPPSS Board of Directors are hoping that the plants will find useful lives as reactors used to burn excess weapons-grade plutonium (*Nuclear News*, June, 1994:20):

> WPPSS has proposed a plan in which WNP-1 and the operating WNP-2, both at Hanford, Washington, would be converted to a mixed-oxide fuel of uranium and plutonium in order to dispose of the weapons-grade plutonium as well as to generate power.

Unfortunately, the *actual* history of the U.S. Nuclear Project, including the WPPSS fiasco, seems to have had little effect on government, military, scientific and corporate support for the technology. Indeed, support for nuclear power seems virtually immune to the negatives of its own history. For example, after the oil crises of the 1970s, nuclear power was frequently cited by its advocates as the path to energy independence while preserving American ideals of abundance. Thus, Harold Agnew, former director of the Los Alamos National Laboratory, declared ten years of escalating fossil fuel prices proved that "nuclear is the only nonfossil fuel energy source that will be available to us in sufficient amounts to supply our current civilization and to fuel progress for the foreseeable future" (1983: 1). The U.S. Department of Energy (DOE) kept faith with the nuclear ideal, even after the Three Mile Island accident in 1978, protecting nuclear R&D throughout the 1980s while cutting conservation and renewable energy funding drastically. In fact, a new program for nuclear energy was initiated with an investment of over $160 million to search for "inherently safe" reactor designs. Westinghouse and General Electric added $70 million of their own money to keep the dream alive (Greenwald, 1991: 61). To date, the American government-corporate partnership has spun off 12 new reactor types for commercialization (*Nuclear News*, September 1992).

Environmental concerns of the 1980s and early 1990s have also been cited to garner support for the U.S. Nuclear Project. Alvin Weinberg has emphasized the nonpolluting nature of nuclear power in his call for a "second nuclear era" (1985). More recently, the National Academy of Sciences released a study supporting rapid deployment of a new generation of "inherently safe" plants to combat the "greenhouse" effect (*Time* Magazine, April 29, 1991: 54). The depiction of nuclear power as "environmentally friendly" is truly remarkable, given the fact that the technology produces some of the most toxic, long-lived and life-threatening wastes known to humankind.

Passage of the 1992 Energy Policy Act (EPACT) rewarded America's nuclear faithful for their persistence. The bill included: $100 million of new funding "for inherently safe" reactor designs;

limits on utility payments for nuclear plant decommissioning with cost recovery from ratepayers; and delegated authority to set high-level waste disposal standards to the National Academy Science in lieu of public participation in standard-setting proceedings. Most important, the Act established a nuclear plant siting system with only one evidentiary hearing prior to the start of construction and it authorized, in principle, advance certification of plant design by the Nuclear Regulatory Commission (see Greenberg, this volume). One industry publication characterized the passage of EPACT as a watershed event in the resuscitation of a moribund industry (*Nuclear News*, November 1992: 34):

> In a single stroke, the [American] nuclear community has now been given virtually everything it has said it needed from the federal government to make nuclear power an attractive, economic choice.

There has been, to be sure, some retreat from the nuclear alternative by the Clinton administration since the passage of the 1992 Act. Nonetheless, the 1995 fiscal year budget calls for over $1 billion of direct and indirect support for the development and maintenance of the U.S. Nuclear Project. This includes some $209 million for civilian reactor development, an additional $122 million for other power technologies, and over $500 million for the Civilian Radioactive Waste Management program, a 40 percent increase over the previous year's appropriations. Despite such handsome subsidies, nuclear supporters find the Clinton Administration's commitment to the technology to be too small (*Nuclear News*, March, 1994: 25).

Enthusiasm for the technology, and belief in its promise, remains high within the ranks of the faithful. For instance, J. Bennett Johnston, senior U.S. Senator from Louisiana, and a long-time and enthusiastic supporter of the technology, continues to actively promote its role in the U.S. energy mix. In a recent interview, Johnston argued that nuclear waste, for the most part, is not "a daunting scientific or engineering problem. [I]t is instead a political problem and to some extent an emotional problem" (*Nuclear News*, November 1993: 47). A number of proposals to prop up the technology continue to circulate through the American political system . These include: a proposal to spend several billion dollars for the International Thermonuclear Experimental Reactor; the implementation of a "one-step" licensing process to keep local or state agencies from obstructing site characterizations at proposed waste sites; and the use of off-budget accounts for federal expenditures associated with DOE's obligations to accept responsibility for high-level nuclear waste generated by civilian reactors (*Nuclear News*, November, 1993: 46-49.)

Federal regulation, as it has done since the beginning of the Nuclear Age, also continues to do its part to keep the technology alive. Thus, in what is only the latest in a long series of accommodating initiatives, the Nuclear Regulatory Commission (NRC) is studying ways to alleviate what the industry sees as an unnecessarily burdensome process required for relicensing. Undaunted by persistent incidences of cracked reactor cores, leaking piping systems requiring early replacement, and numerous other failures, the NRC is preparing to make relicensing a less demanding process (*Nuclear News*, September, 1994: 24-25). At the present time, the rules governing the process, found in 10 CFR 54, require that (*Nuclear News*, September, 1993: 21):

> [A]n applicant for renewal of a power reactor for 10 to 20 years beyond the license's 40-year term must compile the plant's current licensing basis and show how the plant's key systems, structures and components would perform at the 40-year mark and beyond.

The Commission indicated sympathy for the industry's "plight" and has announced its intent to clear away burdens imposed by regulations that are perceived, by the industry, as "uncertain, unstable, or not clearly defined" (*Nuclear News*, September, 1993: 21).

Nuclear Power and the Industrial Order

While an initiator of the nuclear ideal, the U.S. has been joined by most of its industrial allies in the promotion of the technology. France and Japan are among nuclear's most fervent advocates. As well, the former Soviet Union, despite major political differences with the U.S. and other industrialized countries, actively pursued its own Nuclear Project and became a key user and salesman of the technology. With the conclusion of the Cold War, Russia remains a staunch supporter of the nuclear ideal.

The French Experience

By comparison with the U.S., France has encountered fewer obstacles in its pursuit of the nuclear ideal of progress. Nuclear power now accounts for over 70 percent of national electricity production, highest in the world (*Nuclear News*, May, 1992: 53). The country is a leader in nuclear sophistication with its program to commercialize the largest (1,455 MWe) reactor in history (*Nuclear News*, November 1992: 90) and its vitrification technology, designed to secure and store high-level radioactive waste, is regarded as among the most advanced in the world. Perhaps more than any other industrial country, France appreciates the institutional and political requirements for successful growth of nuclear power supply. Indeed, French advocates have even developed an aesthetic vision of nuclear technology as art to bolster support for the technology. Leclerq captures this unusual idea of nuclear power in his comparison of the nuclear cooling tower to some of the grandest architectural monuments of Western culture, including the Arc de Triumph and the Eiffel Tower (1986: 182):

The age in which we live has, for the public, been marked by the nuclear engineer and the gigantic edifices he has erected. For builders and visitors alike, nuclear power plants will be considered the cathedrals of the twentieth century. Their syncretism mingles the conscious and the unconscious, religious fulfillment and industrial achievement, the limitations of uses of materials and boundless artistic inspiration, utopia come true and the continued search for harmony.

For the modernist, France represents the closest Western society has come to realizing the nuclear dream (Rippon, 1992: 86):

The great promise of what might be achieved by the peaceful use of nuclear reactors has largely been realized in France. An advanced industrial country with scarcely any indigenous energy resources, France is today endowed with an abundant and reliable source of clean energy. The French nuclear industry has had its share of problems and no doubt will continue to do so, but with its construction program of the '70s and '80s, it has demonstrated that nuclear power reactors can be built in less than six years, can be commissioned at a rate as high as one unit every two months, and can be operated economically, even in a load-following mode. The French have shown that if there is political will, nuclear energy can make a very large contribution to solving global problems of atmospheric pollution.

As in the case of the United States, French experience with failure has done little to dampen its enthusiasm; indeed, it has often been the case for the French nuclear industry that failure is simply reinterpreted as another sign of success. Thus, a four-year delay in the opening of the 1470-MWe Chooz B1 reactor (originally scheduled for start-up in 1990 and now due to begin operation in 1996), has "not been a total setback for Electricité de France." Rather, the delay is seen positively because it "brought the addition of this large block of new capacity more in line with the load growth of the EdF system," and allowed the plant to be built using more advanced fabricating and operating technologies (*Nuclear News*, July, 1994: 42). Another indicator of the industry's ability to abide failure, was the August 4, 1994 restart of the Super-Phénix fast-breeder reactor. Considered to be a key component of a fully functioning nuclear fuel cycle, the reactor has been plagued by breakdowns, disappearing fuel, and other assorted problems. So serious have these problems been that the reactor has operated only 174 days in eight years (Rothstein, 1994). Nonetheless, a two-year relicensing procedure for the Super-Phénix was completed at the beginning of July, 1994 and the French government has announced that it intends to consider a plan to turn the plant into a plutonium burner rather than a breeder (*Nuclear News*, September, 1994: 90; Rothstein, 1994).

Few critical voices have been heard to challenge the French Nuclear Project. But recent reports suggest that such criticism is, in fact, in order. Mary Byrd Davis, for instance, has documented the extent to which carelessness and mismanagement of waste was a common feature of the French civilian power program (1994). The system is also being scrutinized for its financial practices. According to Rothstein, "the French civilian power industry owes bondholders billions of francs and is increasingly regarded as unreliable" (1994: 8). Nuclear development has also left other parts of the French energy system dangerously exposed. According to one study, "so much money was spent on nuclear power that France neglected to clean up its coal plants. [The result is that] its sulfur dioxide emissions are twice as high per kilowatt-hour as neighboring Germany, which installed scrubbers" (Rothstein, 1994: 9).

Japan and the Plutonium Economy

Japan's embrace of the nuclear ideal may prove to be the most technologically far-reaching. Japan operates the third largest nuclear power system in the world and has already surpassed the U.S. in the percent of national electricity production from nuclear power (24 percent vs 22 percent). In 1993 alone, the country saw the start-up of four reactors with a total generating capacity of 3,799 MWe, increasing Japan's nuclear capacity by 12%. Moreover, Japan has adopted an aggressive construction program to double its nuclear capacity by 2010 (*Nuclear News*, May 1992: 53).

Japan's lack of indigenous energy resources is generally used as the basis upon which to justify its pursuit of a nuclear economy, an argument most recently reprised by the Japanese Atomic Energy Commission in their basic policy statement, the *Long-Term Program for Research, Development and Utilization of Nuclear Energy* (the "Long-Term Plan"). In the country's Long-Term Plan, nuclear energy is portrayed as satisfying two overriding social goals: assurance of a stable supply of energy and the improvement of social welfare (Oyama, 1995: 38). Akira Oyama, vice chairman of the Commission, argues that nuclear energy can be "considered a quasi-domestic energy source produced by technology, making it possible for Japan to overcome its vulnerability in the energy supply system" (1995: 38).

While Japan's aggressive stance toward nuclear expansion is itself noteworthy, it is the nation's commitment to plutonium-fueled fast breeder reactors that sets it apart from all other nations; indeed, plutonium-based technology is considered so hazardous that virtually every other nation (except France) has discarded it as too risky. As Berkhout et al have pointed out (1990: 526):

> The ambition in Japan . . . has been to make the plutonium-fueled fast-breeder reactor [FBR] the

eventual mainstay of the electricity supply system.
Japan's uranium and fossil fuel requirements could
thereby be greatly reduced, bringing freedom from
foreign influence over electricity supplies.

National commitment to this technology has recently been affirmed
in the Long-Term Plan. The Plan argues that uranium, like other
resources, is limited and continued use may cause severe pressures
in supply and demand by the mid-20th century. According to
Oyama (1995: 39):

> It is important, therefore, that [Japan] prepare for
> future energy security by steadily continuing R&D
> efforts towards practical nuclear fuel recycling . . .
> For [plutonium] fast breeder reactors to be
> commercialized by the year 2030, it is necessary to
> pursue R&D towards establishing nuclear fuel
> recycling technology systems of FBRs.

Japan's Long-Term Plan calls for the continued operation of the
Monju and Joyo experimental fast breeder reactors, the
development of a prototype Advanced Thermal Reactor, the
construction of the engineering-scale Recycling Equipment Test
Facility (begun in January 1995), and the full commercialization of
fast breeder reactors by the early part of the 21st century (*Nuclear
News,* July, 1994: 48-49 and June, 1995: 40).

Enthusiasm for the technology is strong in both the public
and private sectors. The political leadership has been (and is)
willing to risk international criticism by shipping plutonium from
France on a seven-week voyage to its Tokai reprocessing plant.
And Satuski Edi, Director General of the Science and Technology
Agency and head of the Japanese Atomic Energy Commission, has
urged the government to augment its reprocessing capacity
(*Nuclear News,* December, 1993: 76). Edi was joined in his
assessment by the heads of Japan's 10 leading power companies,

who put their names to a series of full-page advertisements in British newspapers, which proclaimed, "We don't just support plutonium recycling. We need it." As a demonstration of their commitment, these Japanese business and political leaders pleaded with the British government to give the go-ahead to the Thermal Oxide Reprocessing Plant (THORP) at Sellafield as soon as possible so that Japan could contract for its use (*Nuclear News*, December, 1993: 76).

Notwithstanding such support, early indications are that Japan's ambitious plans might well be endangered by the operating problems that have beset plutonium-breeder reactors. In December 1995, it was reported that the $5.9 billion Monju reactor was shut down after two to three tons of radioactive liquid sodium leaked and began to burn. The plant was reported running at 40 percent capacity when the sodium leaked from a secondary cooling system. The accident was characterized by the government's Nuclear Safety Commission as very serious (*New York Times*, December 17, 1995: A4).

Russia, the Commonwealth of Independent States, and Eastern Europe

Nuclear enthusiasm long ago transcended what, at least at one time, was thought to be the most fundamental ideological division among industrial societies, namely, the contest between capitalism and socialism. The former Soviet Union and its East European allies were among the most bullish nuclear promoters over the technology's 50-year history, operating within their national borders 46 plants representing 10 percent of total world capacity. After the breakup of the Union, Russia still maintains the world's fifth largest nuclear electrical power system, with 25 plants and a rated capacity of 19,800 MWe (*Nuclear News*, March, 1995: 27-42).

Even the worst plant accident in human experience at Chernobyl No. 4 failed to deter Soviet development efforts. A few months after the 1986 Chernobyl explosion, then General Secretary Mikhail Gorbachev assured the world that socialist enthusiasm would not diminish: "The future of the world economy can hardly be imagined without the development of nuclear power . . . [H]umankind derives considerable benefit from atoms for peace" (Vital Speeches of the Day, 1986: 516). Five years later, Soviet Minister for Atomic Energy Vitaly Konovalov announced that the country was committed to expanding its Nuclear Project with 7 GWe of new nuclear capacity to be brought on line by 1995 and an additional 12.6 GWe planned for start-up by the year 2000. In response to a question about the effects of the Chernobyl accident on Soviet thinking, he observed (*Nuclear News*, July 1991: 89):

> [I]n many regions of the country recently there has been a trend, especially among decision-makers and legislators, toward understanding how necessary atomic energy is.

Shortly after Minister Konovalov's statement, the Soviet Union devolved into 15 separate nations. However, the dissolution of the USSR has done nothing to alter commitment to the nuclear ideal. Ending the moratorium against new plant construction imposed after the 1986 Chernobyl accident, the Russian government has recently approved a vigorous program of nuclear power plant construction (see Marples, this volume). The first step towards the realization of this new nuclear capacity was taken with the opening of the Balakovo-4 plant in 1993 (*Nuclear News*, March 1995: 34).

Many of the former republics, as well as satellite countries of the Warsaw Pact, have also maintained the nuclear faith. As in the case of the United States, France, and Japan, these countries' commitments to nuclear power are founded upon promises of economic growth and technological power. The argument seems to be working: the Ukraine announced at the end of 1993 that it will open three new reactors, Zaporozhye-6, Rovno-4, and South Ukraine-4. And while the Ukrainian government has indicated a willingness to shut down the Chernobyl complex, its leaders have also cited a lack of both money for the shutdown and replacement energy as reasons why they must keep the plant running into the foreseeable future (*Nuclear News,* November, 1994: 41).

The same arguments are being repeated throughout Eastern Europe (Hinrichsen, 1993: 37):

> [F]or the troubled nations of the region, the need for electricity is taking precedence over public demands that unsafe plants be closed down permanently. Lithuania's Ignalia plant, though condemned by Western experts, is likely to remain on stream because it produces about 60 percent of the country's electricity. Similarly, Sosnovyi Bor generates 60 percent of the electricity for St. Petersburg; Kozludoy produces 40 percent of Bulgaria's electricity; [and] Paks 40 percent of Hungary's.

The appeal persists despite the well-known economic and technological problems being experienced in the region. Romania, for instance, is readying for start-up of its first ever nuclear plant, the 700 MWe CANDU reactor at Cernavoda. Romania is also considering follow-through on a second unit at Cernavoda, which is currently 32% complete, if it can arrange funding through international partners (*Nuclear News,* October, 1994: 17).

Explaining Nuclear Faith

Why have so many countries, many of them characterized by historically divergent social and economic systems, been so willing to ignore the many failures, risks, and profound dangers associated with the Nuclear Project? In our view, nuclear power represents a logical step in the progression and development of technological society (Ellul, 1964; Mumford, 1934). The Nuclear Project embodies all of the essential elements of the industrial dream — material abundance, technological acumen and independence from the constraints of nature. It also epitomizes the modernist values of scientific rigor, precision and complexity. Indeed, for its supporters, the choice of nuclear power signals the embrace of the modern way of life under the guidance and protection of technological culture. All that stands in the way is pre-technological culture with its "backward" thinking (Lilienthal, 1949: 147-148):

> Atomic energy is a force as fundamental to life as the force of the sun, the force of gravity, the forces of magnetism . . . Within the atomic nucleus are those deep forces, so terribly destructive if used for warfare, so beneficent if used to search out the cause and cure of disease, so almost magical in their ability to pierce the veil of life's secrets . . . For the citizens of the world's leading democracy to be in the dark as to the nature of the fundamental structure and forces of the atom — and of the great good as well as evil this knowledge can bring — would be for them to live in a world in which they are, in elementary knowledge, quite blind and unseeing. It would be almost as if they did not know that fire is hot, that water is wet; as if they did not know there are seasons and gravity and magnetism and electricity.

The attainment of advanced status hinges, in this view, on a future-tense understanding of progress in which enhanced scale, quality and sophistication of technological infrastructure are always valued for their promise of success and present-tense actualities of failure are swept aside. Any other basis of social evaluation is to be judged anti-progressive from this perspective. Langdon Winner has depicted the technological positivism of modernity in this way (1977: 102):

> Certain technical means stand at the very basis of human survival. Failure to provide for them is to invite discomfort, suffering, or even death . . . Any attempt to deny this . . . can only be an expression of malice, stupidity or madness.

The attraction of the industrial world to nuclear power manifests precisely this ideology.

Ideological Transfer:
Nuclear Power in the Third World

> And then there's the question of development: If you have no power, there is no development (M. A. Khan, past chairman of the Pakistan Atomic Energy Commission, quoted in Taylor, 1990: 39).

The transfer of nuclear ideology to the Third World is now underway: over half (23 of 45) of all firm orders for new nuclear plants scheduled for commercial start-up in the 1990s are from developing countries; the remaining 21 plants currently in some phase of construction are located in Russia or the CIS countries (*Nuclear News*, March, 1995: 27-42). The rationale for nuclear power in the Third World, as with the "advanced" tier, has had little to do with the present-tense conditions and needs of societies.

The elite of the Third World have been courted for over three decades to provide leadership in diffusing nuclear technology and values throughout the region. This elite has been educated in the ideology by the International Atomic Energy Agency (IAEA) and other multilateral organizations and has, in turn, supplied them with some of their recent leaders. A prime example, in this regard, is Munir Ahmad Khan, who served as chairman of Pakistan's

Atomic Energy Commission and then took over the reins of the IAEA. He is one of the world's strongest advocates of this highly expensive, esoteric technology. He makes the case for nuclear power in the Third World on familiar grounds. First, Khan defines the issue in terms that echo the long-held Western belief in increased energy supply as a prerequisite for the advance of civilizations (see Basalla, 1980). Khan argues that (1992: 76):

> The developing countries desperately need electric power to speed up their industrialization and improve their economic lot, to overcome poverty and forestall the social and political upheavals that have rocked Eastern Europe.

The issue can be framed in precise terms:

> [Developing] countries constitute about 67 percent of the world population, but consume only 17 percent of world energy. The average annual per capita electricity consumption stands at 0.7 megawatt-hours in the developing countries, versus 6.5 MWH in industrialized countries. In addition to having low electricity consumption, these countries are also deficient in conventional energy resources. Excluding the few oil-rich countries, the per-capita energy reserves in the developing countries amount to less than 45 tons of oil equivalent, compared to 366 tons in the industrialized countries.

This leads Khan to the nuclear solution:

> This is why the energy-starved developing countries look to nuclear power as a potential source of meeting their future electricity needs at a reasonable cost and reducing the increasing burden on their debt-ridden and fragile economies. If the

nuclear power alternative is not available to
[developing countries] for technical, financial, or
political reasons, they will inevitably turn toward
using oil or even poor-quality coal, which will
greatly increase carbon dioxide emissions to the
atmosphere.

Seen in these terms, nuclear power is "a practical choice which is
economic, less polluting, more reliable, and affords . . .
diversification" (Khan quoted in Taylor, 1990: 38).

Khan's assessment contains many of the future-tense
arguments made in the U.S. and elsewhere to promote the
development of nuclear technology. But the appropriateness of
these arguments for the Third World are doubtful. Developing
countries typically lack the necessary investment capital, research
and technical infrastructure, and fully articulated electric grids to
"plug in" a nuclear plant. Moreover, access to energy services is
often more important in defining social need than the amount of
available supply. And surely, developing countries should not
assume the burden of reducing greenhouse gas emissions at this
time. Quite the reverse, it is the low-carbon development pattern
of the Third World that has, so far, offset the overuse of the
atmosphere by industrial countries for storing CO_2 (see Byrne et al,
1994).

To date, present tense objections to nuclear power have
been no more successful in the Third World than in the industrial
tier. Developing countries with nuclear power aspirations routinely
justify their interest on the basis of future economic and
technological benefits. Actual social problems and costs are made
abstract while unrealized possibilities are treated as though they
were concrete. This logic has found favor throughout Latin
America, Asia and the Indian subcontinent. The cases of Mexico,
Argentina, Brazil, South Korea, Taiwan, India, Pakistan and China
are instructive for the revealing glimpse they provide into Third

World elite thinking on the relations between energy, technology and development.

Latin America

Mexico's experience is illustrative of Latin countries that have sought national progress and energy independence through atomic development. As a leader in the movement to control global proliferation of nuclear weaponry, Mexico initially showed little interest in the peaceful use of the atom. By the mid-1960s, however, the country was in the process of institutionalizing the capacity to build and operate indigenously fabricated nuclear power plants, based upon a long term goal of "infus[ing] knowledge and skills that could then be applied to future development" (Stevis and Mumme, 1991: 60). Mexico pursued this policy despite the fact that it has an abundance of conventional fossil fuels, including proven and potential oil reserves currently estimated at about two trillion barrels (Miramontes, 1989: 36). Key government officials and scientists from the National Council for Science and Technology, the Physics Institute of the National University, and the Institute for Nuclear Research persuaded the government that nuclear power was essential for Mexico "because it signals an era of progress and modernity" (Miramontes, 1989: 38). These beliefs have not been shaken by the experience of steep present-tense costs at the country's experience 654 MWe Laguna Verde nuclear reactor. As Miramontes points out (1989: 38):

> Nearly all of the nuclear technology has been imported. In turn, Mexico gets hard currency from oil exports. This means that Mexico must sell hydrocarbons to pay for the nuclear power plant. It has been estimated that Mexico will have to export about 345 million barrels of oil to pay for the [Laguna Verde] plant, but the plant will save only about 240 million barrels. In an effort to save

hydrocarbons, Mexico will lose 105 million barrels of oil and gain hundreds of tons of radioactive wastes.

Like Mexico, Argentina and Brazil pursued nuclear power for future-tense reasons. A healthy dose of military aspirations also attracted the two countries to nuclear energy. According to Adler, both countries' nuclear ventures can best be understood by taking into account the ideology of autonomous development and industrial development (1988). The quest for "nuclear autonomy," according to Adler, is part of a larger quest to achieve international parity (1987:18):

> [P]rogress is viewed . . . not only as modernization and economic and technological development but as a matter of autonomy and equality as well. This is why [each country's] nationalist ideology is so strongly linked to development and equality. Liberation, cultural self-affirmation, development, science and technology: these are the core dimensions of the idea of progress in the Third World.

Brazil recently announced its intent to continue its program, and specifically to complete the Angra-2 plant. Despite having spent some $4.6 billion on Angra-2 and 3, the plant (which was originally begun in 1976) it is still only 69% complete. The President is recommending that an additional $1.4 billion be spent in order to bring the plant into operation (*Nuclear News*, October, 1994: 59).

Asia's Developing Countries

The Korean Peninsula. Asian developing countries have demonstrated a keen interest in nuclear power not only in their development plans, but in a willingness to, in the vernacular, "pour the concrete." Two of the "Asian Tigers" — South Korea and Taiwan — have already installed 7.2 GWe and 4.9 GWe of nuclear capacity, respectively, ranking them 9th and 12th in the world. Moreover, the two countries have adopted the world's most ambitious nuclear expansion plans for the 1990s; Korea alone accounts for over 10 percent of all new orders placed to date for start-up in this decade (*Nuclear News*, March, 1995: 27-42).

Perhaps the most impressive of any nation's commitment to a nuclear future is South Korea's. The country currently has nine plants in operation and plans to add nine more by the year 2001, as well as an additional nine plants by 2006. It is currently the most nuclear-intensive country in the developing world (Taylor, November, 1992: 41). South Korea's nuclear capacity places it ahead of Spain, Belgium, Bulgaria, Hungary and Finland and, in percent of electricity generation supplied by nuclear power, it outpaces Germany and Japan, as well (*Nuclear News*, May 1992: 53 and August, 1992: 55-72). When the 18 new plants come on line by 2006, nuclear power will be supplying well over half of the nation's total electricity needs.

Conditioned by the need to suddenly create a free-standing electrical generation system as a result of North Korea's cutoff of its electricity in 1948, South Korea has consistently emphasized energy abundance in its development strategy (Kim and Byrne, 1991). The centrality of nuclear power, in the leadership's view, in lifting South Korea out of its impoverishment after civil war was articulated at ground breaking ceremonies for Kori-1, the country's first commercial nuclear reactor (556 MWe). The late president of South Korea, Park Chung-Hee, spelled out the Korean version of the energy-civilization equation (1971: 144):

> We are very proud of and happy that this country
> is constructing the most technologically-advanced
> [nuclear power plant] in the late 20th century. As
> we realize, electricity is what all countries of the
> world want for economic development . . . Until

now we have not shared in enough benefits of electricity in this country . . . Awareness of this fact would let us understand how important the promotion of electricity generation is and, by constructing many nuclear power plants, how much benefit from electricity we can we receive. Furthermore, it will be possible that we can advance the larger economic development and lead the country to a higher cultural life.

Recent forecasts of electricity requirements by the Korea Electric Power Company (KEPCO) are evidence of South Korea's desire to bring the country's current average annual electricity consumption per capita of 2,500 kWh up to the 10,000 kWh consumed in the United States by early next century (Taylor, November, 1992). South Korean energy officials take special pride in the better than 10 percent average annual increases in electricity demand that have occurred over the last twenty years. While the rate of increase slowed somewhat in recent years, KEPCO officials are convinced that sizable consumption growth will continue and, for this reason, intend to stay the course of rapid expansion of the country's electrical network. In their view, nuclear power is the only viable option for an energy-poor country with high demand growth. In this vein, KEPCO continues to cite favorably a proposal by leading scientists and energy researchers of the country to construct 50 additional nuclear plants by 2031 (KEPCO, 1989). One nuclear industry commentator has summarized the country's self-rationalization of its extraordinary commitment to nuclear power (Taylor, November, 1992: 41):

[I]n a region where underdeveloped third-world countries have economies crucially hobbled by inadequate power generation, South Korea has avoided that pitfall by staying ahead of the curve in providing adequate electricity generating capacity to fuel its astoundingly burgeoning economy.

New to the list of "nuclear hopefuls" is North Korea. While the political and social isolation of this country makes it difficult to obtain reliable information on its activities, there is a reasonable basis for believing that North Korea will soon become home to at least two 1,000 MWe plants. There are a host of problems to overcome, including financing, the yet-to-be demonstrated capacity of North Korea to build, maintain and operate the plants, and South Korean willingness to supply the necessary LWR technology.   Still,   many   commentators   predict   eventual construction of the plants (*Nuclear News*, November 1994: 41).

A particularly intriguing aspect of the North Korean case is that international negotiations, led by the U.S. and Japan, have advertised the installation of these plants as a step toward bringing this country into the mainstream, and at the same time, contributing to peace in the region. In the North Korean nuclear bargain, the special language, thinking, and values of the atomic age are displayed. What clearer indicator can there be of the extraordinary power of nuclear ideology than the association of peace and normalcy with the transfer of the world's most dangerous technology to what many regard as a rogue nation?

India and Pakistan. Nations of the Indian subcontinent are also firmly committed to nuclear expansion. India currently operates 9 plants with a capacity of just over 1600 MWe. The country is planning to have at least 10,000 MWe of installed nuclear capacity by the year 2000. Currently, Pakistan has only one small 125 MWe nuclear plant. But it, too, has actively sought to build a nuclear future. As with the Korean peninsula, the Indian subcontinent's pursuit of nuclear power has raised concerns about the technology's pursuit for military aims in a region plagued by armed conflict. Yet, here too nuclear power is held out by advocates as a source of hope for peace.

India's choice of nuclear power was made by its leaders in the name of sovereignty, independence and international status.

The country has remained steadfast in its commitment to indigenous nuclear technology design and construction, notwithstanding the daunting challenges of poverty facing the society.

From the outset, India conceived its national nuclear program as key to confirming the country's arrival in the modern era. The architects of the Indian Revolution accepted early on the nuclear maxim that energy abundance is the foundation of social progress. Nehru argued in 1948 that India had missed the first Industrial Revolution due to her lack of technical skill and that success in the Second Revolution hinged upon the nation's development of a nuclear energy program. Two decades later, Indira Gandhi characterized nuclear power as an essential technology necessary for rescuing developing nations "from the shackles of poverty and ignorance" (quoted in Pathak, 1980: 24-25). In a national speech, Prime Minister Gandhi presented a vision of energy-intensive development that is hardly distinguishable from the technocratic model widespread in the West (quoted in Pathak, 1980: 24-25):

> Our programme of atomic energy development for peaceful purposes is related to the real needs of our economy and would be effectively geared to this end. Atomic energy stations would play a valuable role in the future not only in areas where other sources of energy are expensive but as base-load stations working alongside large hydro-electric installations. The significance of all this to our economy which is so heavily dependent on agriculture is tremendous.

The country's energy planners continue to regard nuclear power as the "ultimate dream" (*India Today*, 1988: 87) for supplying electricity in amounts needed for India's economic development. Toward this end, India has pursued the development of all phases of the nuclear system. Work continues, for instance, on two reactors at Kaiga in Karnataka despite a 1994 construction site accident. Two more units of the 235 MWe design are under construction (Rajasthan-3 and -4) and another four units are planned (Rajasthan-5 to -8). India has also developed the technology for mixed plutonium/uranium oxide fuel fabrication, has sufficient milling facilities to meet expected requirements through the early years of the next century, and has developed facilities for the production of zirconium and the fabrication of zirconium alloy tubing, both for fuel cladding and for the calandria tubes required in PWR-type reactors. India has also operated heavy-water production plants as well as a number of small reprocessing facilities since at least the late 1960s (Rippon, 1995).

While Pakistan's nuclear involvement is currently limited to one operating plant, is also planning significant new investments in nuclear energy. Perhaps most important, the country has been "working hard to develop an indigenous capability to design and construct a series of standardized nuclear power plants" (Rippon, 1995: 42). In the meantime, according to Rippon, "efforts have been made to find international vendors to supply the nuclear plants that are clearly needed to help meet the growth in energy demand" (1995: 42). Thus, in November 1989, the country announced that it had agreed to purchase a 300 MWe pressurized water reactor based on the Chinese-designed plant in commercial service at Qinshan. The contract covered both the supply of the nuclear power plant with its fuel and the transfer of technology and other support services. Construction commenced at the Chasma site in August 1993, with a scheduled start-up date of August, 1998. The country's hydro-thermal power program calls for a total nuclear capacity of 4,625 MWe by the year 2006 (Taylor, March, 1990: 38).

China. Finally, China's bid for recognition by the community of nuclear states exemplifies the pervasiveness of the technology's ideological appeal in the Third World. While China's rulers have sought to put the country on a development road of its own definition and making and have, until recently, set restrictions on economic and technological contact with the West, its nuclear program has invited participation from American, French, British, German and Japanese corporations. Moreover, the country's strategy is modeled essentially along the same lines as ones in the industrial countries.

Qinshan-1, the country's first nuclear facility, reached criticality October, 1991 (*Nuclear News*, March, 1995: 29) while two other 900 MWe reactors at Guangdong's Daya Bay (near Hong Kong) began commercial operation in 1993 (*Nuclear News*, September, 1994: 89). According to an industry commentator, two factors are driving the Chinese nuclear program: "the need to boost its technological prowess, providing new products for Chinese exporters;" and "national prestige . . . [which is] motivating mainland China's push to become a full-fledged player in the nuclear power game" (Gallagher, 1990: 106-107). Chinese leaders have also indicated their acceptance of the cardinal principle of future-tense evaluation of nuclear industrialization. In terms reminiscent of the Soviet official response to the Chernobyl explosion, former premier Chao Tzu-yang acknowledged that the technology's history of accidents had forced greater attention to issues of safety. But, he pointed out, "that will not change our attitude toward developing the nuclear power industry" (Gallagher, 1990: 109). The reality of risk is discounted by the unrealized possibilities of nuclear-inspired development.

Many observers also believe that the recent opening of plants at Qinshan and Guangdong "has stimulated a new sense of confidence in the emerging nuclear industry, and [that] there is reason to believe that the latest predictions of an imminent take-off may soon be realized" (Rippon, June, 1995: 32). In the immediate future, plans call for the addition of two 600 MWe units at Qinshan, as well as two more 950 MWe units at Lingao, which is adjacent to Guangdong. Also, several of the more prosperous provinces along the eastern seaboard are actively planning nuclear projects, which, if they all come to fruition, will add an additional 20 GWe of nuclear capacity in the first decade of the next century. An additional 50 GWe of capacity is called for in the more ambitious plans of the central authorities (Rippon, June, 1995: 32).

As in the case of other countries in the region, China is also aggressively pursuing an indigenization policy.   However, the nation's leaders are not allowing the absence of a completely developed domestic nuclear system to delay their plans. Thus, in recent years, China has signed agreements with a full set of international partners. The Russian-based St. Petersburg Atomic Energy Design Company, for example, is now working with the China National Nuclear Corporation (CNNC) to design two 1,000 MWe units at the northeastern province of Liaoning. In February of 1995, South Korea concluded a Memorandum of Understanding (MOU) for the supply of two units of its so-called Korean standardized nuclear power plant. The South Koreans are also working with CNNC on an assessment of prospective sites in Shandong and Fujian provinces. Finally, Atomic Energy of Canada Limited has also concluded an MOU designed to facilitate the sale of its CANDU reactors to China. According to one commentator, the "move exemplifies . . . the ascendency of the Asian market as power demand grows in developing Pacific Rim countries" (Rippon, June, 1995:33)

The reliance upon international vendors is likely to be only temporary since China is fairly far along in achieving its goal of indigenization. To this end, the country has invested in all stages of the nuclear fuel cycle, including the development of extraction and refining, fuel processing, and commercial fuel fabrication facilities. China also is experimenting with reprocessing activities and has initiated a pilot-scale reprocessing plant at Lanzhou. Finally, China is developing waste management programs, including low- and intermediate-level repositories, as well as deep-site geological storage capacity (Rippon, June, 1995: *Nuclear News,* November, 1994: 40 and February, 1994: 54).

The Lure of the Nuclear Dream

As in the West, the lure of nuclear power in the Third World derives from shared ideas of technological success as key to social progress. Countries with nuclear ambitions equate energy use with civilization, material abundance with national independence and technological sophistication with social progress. Military aspirations are, of course, a part of the equation, as well. But this only underscores the thinness of the distinction between civilian and military nuclear programs, a feature that advocates generally prefer not to discuss.

There is an irony in the shared aspirations of the West and the Third World toward nuclear power. Nations that are otherwise related by the contradiction of extravagant wealth amid desperate poverty have mutually embraced an ideology that presumes a general condition of harmony between them on matters of technology and development. Indeed, as Adler (1988) points out, on the question of nuclear energy, developing countries often assume that the technology is a force for parity and, therefore, *if anything* the Third World should be wary of possible industrial country efforts to prevent its full utilization. The adoption of this view assures that nuclear power is exempted from present-tense social criticism and results in the Third World being a participant in its own exploitation. In this way, ideas of technology and development that rationalize industrial hegemony — what Jacques Ellul (1964) termed the vanguard of "technical invasion" — come to inform the aspirations of the leadership of the exploited.

Conclusion

Finally, as the 20th century draws to a close, the perseverance of the Nuclear Dream warns of the era's near-complete failure to break through the facade of technological progress. It remains at least as difficult at the end of the era of the first technological century as at its beginning to recognize and value the *actual, present-tense lifeworld* ahead of the world of technique. Beyond its record of secrecy, contamination, financial boondoggle, catastrophe, and near-catastrophe, the Global Nuclear Project stands as stark testimony to the era's willingness to deny the authoritarian reality that has universally accompanied the technology's development in favor of its promise of future-tense abundance. The lifeworld risked for the ideal of More — this is perhaps the most disturbing legacy of nuclear power.

References

Adato, Michelle, et al. 1987. *Safety Second: The NRC and America's Nuclear Power Plants.* Bloomington, IN: Indiana University Press.

Agnew, Harold B. 1983. "Civilian Uses of Nuclear Power: Status and Future." *Proceedings of the Symposium on Energy: Challenges and Opportunities for the Middle Atlantic States.* Volume 2. John Hopkins University. Energy Research Institute. Baltimore, MD.

Adler, Emanuel. 1988. "State Institutions, Ideology, and Autonomous Technological Development: Computers and Nuclear Energy in Argentina and Brazil." *Latin American Research Review.* Volume 23, No. 2: 59-90.

_____. 1987. *The Power of Ideology: The Quest for Technology Autonomy in Argentina and Brazil.* Berkeley, CA: The University of California Press.

Basalla, George. 1980. "Energy and Civilization." In P. Starr and P.C. Ritterbush (eds). *Science, Technology, and the Human Prospect.* New York, NY: Pergamon Press.

Berkhout, Frans, Tatsujiro Suzuki, and William Walker. 1990. "The Approaching Plutonium Surplus: A Japanese/ European Predicament." *International Affairs.* Volume 66, No. 3: 523-543.

Byrne, John, Constantine Hadjilambrinos, and Subodh Wagle. 1994. "Distributing Costs of Global Climate Change." *IEEE Technology and Society.* (Spring): 17-32.

Byrne, John and Steven M. Hoffman. 1992. "Nuclear Optimism and the Technological Imperative: A Study of the Pacific Northwest Electric Network." *Bulletin of Science, Technology and Society.* Volume 11: 63-77.

_____. 1988. "Nuclear Power and Technological Authoritarianism." *Bulletin of Science, Technology and Society.* Volume 7: 658-671.

Byrne, John, Steven M. Hoffman and Cecilia R. Martinez. 1992. "Environmental Commodification and the Industrialization of Native American Lands." *Proceedings of the 7th Annual Meeting of the National Association for Science, Technology and Society:* 170-181.

_____. 1989. "Technological Politics in the Nuclear Age." *Bulletin of Science, Technology and Society.* Volume 8: 580-594.

Byrne, John and Daniel Rich. 1986. "In Search of the Abundant Energy Machine." In John Byrne and Daniel Rich, eds. *The Politics of Energy Research and Development.* Energy Policy Studies, Volume 3.New Brunswick, NJ: Transaction Books. Pp.141-160.

Camilleri, Joseph A. 1984. *The State and Nuclear Power: Conflict and Control in the Western World.* Norfolk, UK: Wheatsheaf Books, Ltd.

Carter, Luther. 1987. *Nuclear Imperatives and the Public Trust.* Washington, D.C.: Resources for the Future.

Comptroller General of the United States. 1984. *Status of BPA's Efforts to Improve its Oversight of Three Nuclear Power Projects.* Washington, D.C.: U.S. Government Printing Office.

Davis, Mary Byrd. 1994. "The French Mess Nucléaire." *Bulletin of Atomic Scientists.* (July/August):48-53.

Ellul, Jacques. 1964. *The Technological Society.* New York, NY: Vintage Books.

Gallagher, Michael. 1990. "Nuclear Power and Mainland China's Energy Future." *Issues and* Studies. Volume 26, No. 12: 100-120.

Greenwald, John. 1991. "Time to Choose." *Time Magazine.* Volume 137, No.17 (April): 54-61.

Hinrichson, Don. 1993. "Russian Roulette: A New Nuclear Threat from Eastern Europe." *Amicus Journal.* Volume 14, No. 4 (Winter): 35-37.

KEPCO. 1989. *The Outlook and Development Strategy of Nuclear Energy for the Early 21st Century in the Republic of Korea.* Seoul, Republic of Korea: KEPCO and A Ju University.

Khan, Munir Ahmad. 1992. "Fifty Years of Energy in the Third World." *Nuclear News.* November: 75-76.

Kim, Jong-dall and John Byrne. 1991. "Centralization, Technicization and Third World Development on the Semi-Periphery." *The Bulletin of Science, Technology and Society.* Volume 10, No.4: 212-222.

LeClercq, Jacques. 1986. *The Nuclear Age.* Paris, France: Le Chene.

Laurence, William L. 1959. *Men and Atoms.* New York, NY: Simon and Schuster.

_____. 1940. "The Atom Gives Up." *Saturday Evening Post.* (September 7): 12-13.

Lilienthal, David E. 1949. *This I Do Believe.* New York, NY: Harper and Row.

Miramontes, Octavio. 1989. "Wooing Mexico to Nuclear Power." *Bulletin of the Atomic Scientists.* Volume 45, No.6 (July/August): 36-38.

Mumford, Lewis. 1934. *Technics and Civilization.* New York, NY: Harcourt, Brace.

Noble, David. 1983. "Present Tense Democracy: Technology's Politics." *Democracy.* Volume 4, No. 2: 8-24.

Northwest Power Planning Council. 1988b. *The Role for Conservation in Least-Cost Planning.* Portland, OR: Northwest Power Planning Council. (June 10).

Nuclear News. "World List of Nuclear Power Plants."; March, 1995: 27-42."Framatome to help build Qinshan-2 and 3." November, 1994: 40; "Ukraine is ready 'in principle'to close Chernobyl." November, 1994: 41; "Commissioning of Romania's first nuclear power plant" October, 1994: 17; "President Franco calls for Angra-2 completion." October, 1994: 59; "Com Ed OK'd to run units with core shroud cracks" and "Leaking valves mar otherwise smooth restart." September, 1994: 24-25; "GNPGC announces plans for 10 more nuclear plants." September, 1994: 89-90;"Super-Phénix rises again after four-year shutdown." September, 1994: 90; "France's first N4 unit at Chooz B nears startup." July, 1994: 42-43; "Revised plutonium program in latest energy plan." July, 1994: 48-49; "WPPSS Voted On May to Terminate WNP-1 and -3." June, 1994: 20; "Nuclear funding cut to boost other energy sources." March, 1994: 25; "1993 ended with 419 nuclear units worldwide." March, 1994: 18; "Nuclear plans expanding, says CNNC Chair Jiang." February: 1994: 54; "Japan sees PU recycle as a responsibility to the future." December, 1993: 76; "Ukraine will open three new reactors and keep Chernobyl." November, 1993: 19; "Johnston: You can't solve the energy problem, but you can cope with it." November, 1993: 46-49; "License Renewal May Become a Less Demanding Process." September, 1993:21; "Congress passes very pronuclear energy bill." November 1992: 33-34; "The New Reactors." September, 1992: 66-90; "World List of Nuclear Power Plants." August, 1992: 55-72; "Nuclear contributions in 1991." May, 1992: 53; "Moscow once again sees nuclear expansion ahead." July, 1991: 89.

Oyama, Akira. 1995. "Japan's Nuclear Future." *Nuclear News.* June: 38-40.

Olson, Daryl. 1982. "The Washington Public Power Supply System: The Story So Far." *Public Utilities Fortnightly.* June 10: 15-26.

Park, Chung-hee. 1971. "Speech of Gratitude: Start of Construction of the First Nuclear Power Plant: Kori-1" (March 19). In *Speeches of President Park Chung-Hee.* Ministry of Culture and Information. Republic of Korea (in Korean): 141-145

Pathak, K. K. 1980. *Nuclear Policy of India: A Third World Perspective.* New Delhi, India: Gitanjali Prakashan.

Rippon, Simon. 1995. "Asian Subcontinent: Nuclear programs in Pakistan, India." *Nuclear News.* June: 42-43.

_____. 1992. "Focusing on today's European nuclear scene." *Nuclear News.* November: 81-100.

Rothstein, Linda. 1994. "French nuclear power loses its punch." *The Bulletin of Atomic Scientists.* (July/August):8-9.

Stevis, Dimitris and Stephen P. Mumme. 1991. "Nuclear Power, Technological Autonomy, and the State in Mexico." *Latin American Research Review.* Volume 26, No.3: 55-82.

Taylor, Gregg M. 1992. "KEPCO plans 18 more reactors by 2006." *Nuclear News.* November: 41-49.

_____. 1992. "Japan: A look at its future." *Nuclear News.* July: 32-37.

_____. 1990. "Pakistan: Energy poor and seeking more nuclear power plants." *Nuclear News.* March: 38-45.

Weinberg, Alvin. 1985. *The Second Nuclear Era: A New Start for Nuclear Power.* New York, NY: Praeger.

_____. 1972. "Social Institutions and Nuclear Energy." *Science.* 177: 27-34.

_____. 1956. "Today's Revolution." *Bulletin of the Atomic Scientists* (October). 299-302.

Winner, Langdon. 1977. *Autonomous Technology.* Cambridge, MA: The MIT Press.

Chapter 2

Nuclear Policy as Projection:
How Policy Choices Can Create
Their Own Justification

James M. Jasper

Introduction

The modern nation state rarely intervenes in the economy or in technological development as extensively as it did with nuclear energy. Perhaps only a technology arising out of military research could gain such support, at every stage of its commercialization and in every facet of its operation–from uranium mining to fuel enrichment to electricity production to disposal of radioactive wastes. In the United States, in particular, such extensive and prolonged intervention is unique outside of war. Exceptional cases can nonetheless offer important lessons, highlighting aspects of policymaking otherwise overlooked. In this case, we can see policymakers' elaborate predictions about the future, and their ability to make predictions — however unrealistic — come true.

A cross-national comparison will also help us focus on the choices made by policymakers in developing nuclear energy. In most political analyses the United States is said to have a "weak state" and France a "strong state," as though states were rigid structures fixed for all time instead of constantly changing institutions that officials — and others, in some cases — can use to further their ends. If the U.S. government intervenes less often or directly than the French, the reason has as much to do with persistent laissez-faire ideologies as any lack of capacities. Elected and unelected officials in the United States intervene less often — and cross-national differences on this dimension are easily exaggerated — more because they don't want to than because they cannot.

A brief look at nuclear development in France and the United States shows the power of officials to reshape the world around them. First through hortatory predictions about future events, then through active management of technological and industrial policy, political, economic, and technological elites heroically and successfully created an entirely new industry, based on a new way of generating electricity.

Policy as Projection

In the advanced industrial nations, the state is involved in many forms of supposedly scientific prediction: economic and revenue forecasting, weather predictions, environmental impact assessments, demographic analyses, estimates of the likely course of diseases. Virtually *every* long-term policy is a form of prediction, involving efforts to predict the future course of events as well as to shape them. This is especially true of policies to develop new technologies. To some extent, such policies alter the "givens" guiding decisionmakers, who "project" onto the world their own image of how it should be, first predicting or claiming that the world looks a certain way, then working to make it actually look that way.

Because of the uncertainties, large engineering projects (both social and technical) frequently turn out more unforeseen consequences than intended ones. If nothing else, the changing contexts in which policies unfold — economic crises, the price of oil, international relations, the outcomes of the next election — mean that the results of many policies cannot be predicted with much confidence. The level of uncertainty is even greater in the case of technological development, which always involves unforeseen twists and obstacles and can never be fully plotted in advance. The point of this kind of policy, after all, is to create something new. Some state policies are even designed to force technological development, for example in the area of environmental protection; goals are selected which cannot be attained with existing technological means.

Like religious rituals or magic, predictions give us a sense of control over essentially uncontrollable forces, apparent knowledge about unknowable outcomes. Because they impose rules and narrow our options, predictions can be a form of *social* control as well, a power that experts use to tell the rest of us what to do. We give discretion to experts because they promise to protect us from dangers, because they claim to know enough to control the natural world (Jasper, 1992a). They use an instrumental rhetoric of "this is how things are, this is the nature of things, it must be this way." This language distinguishes them from the rest of us: it's the power of expertise, the power of special knowledge about the way things are. (In a way, we want — even demand — the reassurance of their predictions. That is why natural disasters are often a threat to regimes in the advanced industrial countries: the experts have reneged on their implicit promise.) Time after time, experts take advantage of their authority, using their predictions in political ways, to further their own goals and increase their own power.

At the same time, policy predictions by state officials and others can be a positive projection of their vision and hopes. Predictions can be fantasies, utopias, even a kind of intoxication (Bupp and Derian, 1978). Predictions show a world in the process of becoming, with a foot in the present and a foot in the future, a kind of existential project of world creation. A prediction begins as a kind of anticipation: a structure, a path, a blueprint that begins in fantasy, purely in the mind. But as soon as it is expressed in the form of a prediction, people can begin filling it in with a more concrete kind of reality, making the vision come to life, giving it an institutional reality — and being held accountable to it.

Enthusiasm for technological progress is, of course, a mainstay of the modern Western world, and now almost the entire world. And at least since the Romantic movement of two hundred years ago there has been considerable ambivalence about it. Today, I think, we tend to judge technological enthusiasm by its democratic accountability, and we tend to judge it rather harshly as a result. In the advanced industrial countries there is widespread ambivalence about science, technology, and experts. What we de-emphasize today is the heroic, creative side of these activities.

Forcing Nuclear Technology

The heroic side of civilian nuclear energy demonstrates a faith in the ability of engineers and scientists to do the most outlandish things, to do virtually whatever they wish, to expand the limits of what we can imagine to be possible, and in fact to expand the limits of what is physically possible. There are only a few other cases as impressive as nuclear energy, such as the airplane or space exploration. The very idea of getting electricity by splitting the atom had an almost magical grip on the imaginations of inventors and policymakers. As soon as someone said — in an even mildly credible way — that these things *could* be done, then people quickly convinced themselves — at least the people involved in the projects — that they *would* be done. Policymaking became almost a collective fantasy, or at least was driven by fantasy.

Energy policy choices should presumably be guided by criteria such as cost, reliability, resource availability, safety and (we can see this better today than planners did in the 1970s) public opinion. All these were used as part of the official rationale for nuclear programs, but in fact very little was known about any of them at the time major commitments were made. What's more, all changed drastically under the impact of the policies themselves.

In the United States the early predictions for nuclear power were downright giddy. One popular magazine article foresaw "a new Eden," based on nuclear energy, with the abolition of "disease and poverty, anxiety and fear" (quoted in Hilgartner et al, 1982: 39). Nuclear electricity would be "too cheap to meter," predicted Atomic Energy Commission chairman Lewis Strauss in 1954. Several years later, Alvin Weinberg proclaimed, in the face of all available evidence, "nuclear reactors now appear to be the cheapest of all sources of energy" (quoted in Bupp, 1979: 136).

What institutional factors led people to make such predictions about nuclear energy, and to believe them? Scientists, according to David Lilienthal (1963), were relieved to find a peaceful use for the research accomplished by the Manhattan Project. Second, even though the military reluctantly ceded control over civilian development, the Atomic Energy Commission (AEC) and Congress' Joint Committee on Atomic Energy retained much of the insulated power of military arrangements. The AEC interpreted its role less as regulation than as promotion of the new technologies, and the Joint Committee was filled with legislators with special interests in technological development who consistently proved more optimistic about the future of the technology than either the AEC or the engineers with hands-on experience. Everyone involved wanted the reactors to work, and persuaded themselves of how cheaply and soon they would be available. Frustrated at the lack of utility interest in what was obviously the technology of the future, the AEC and the Joint Committee offered a series of carrots (subsidies for pilot reactors, accident insurance underwritten by the government, free uranium for seven years) and sticks (the Gore-Holifield bill of 1956 proposed actual government construction and ownership of reactors).   Regulators' main tool, though, was simply loud exhortation and optimism. In a 1962 report, for example, the AEC proclaimed that nuclear energy was "on the threshold of being competitive with conventional power in the highest fuel cost areas," predicting that half the country's electricity would come from nuclear by the year 2000 (Mazuzan and Walker, 1984: 409-418). All this was to little avail, however, until manufacturers began to share these optimistic predictions.

The first companies manufacturing nuclear reactors, primarily GE and Westinghouse, believed they could make a lot of money from this endeavor, so they wanted very much to believe in the enthusiastic predictions. These two companies were directly competitive with each other. Both hired new CEOs in the early 1960s, neither of whom knew much about nuclear energy. One was a former light bulb salesman, the other a financial analyst. They took the predictions as a matter of faith — literally — much more than many of the engineers in the companies. And they acted as though the predictions would come true. With enormous amounts of money at stake in the predicted market for reactors, each company tried to get the jump on the other by selling loss-leaders: "turnkey" plants sold for a price fixed in advance, a price designed to make nuclear energy look competitive with fossil fuels.

These cheap plants (on which the two companies together lost almost one billion dollars) did what no government programs had: convince America's electric utilities that nuclear technology had reached commercial maturity.   From 1965 to 1968, accordingly, 74 new reactors were ordered–all of them larger than any yet in operation. In other words, dozens of utilities (ultimately, more than sixty) committed vast sums of money on the basis of a few turnkey plants (which provided poor evidence about what future plants would cost) and sheer conjecture about the economies of scale expected from ever-larger reactors. Unexpectedly poor results from the few small plants already in operation were dismissed as learning costs, not applicable to the larger reactors. In the words of Nuclear Regulatory Commissioner Peter Bradford, "an entire generation of large plants was designed and built with no relevant operating experience — almost as if the airline industry had gone from piper cubs to jumbo jets in about fifteen years" (1982:10).

With the government cheering from the sidelines, manufacturers and utilities competed with each other in their optimistic predictions. When General Electric sold a reactor to the Tennessee Valley Authority in 1966, it was so sure of its construction speed that it offered to pay $1,500 per hour of peak demand when the reactor was not operating properly. The TVA, in a decision it must have later regretted, matched GE's optimism by turning down the offer (Pringle and Spigelman, 1981: 269). Most utility officials saw nuclear fission as "just another way to boil water," but with a cheaper, cleaner fuel than oil or gas. Mostly engineers, not economists, their faith in technological progress precluded much skepticism about fanciful cost predictions. In this regulated industry, nuclear energy faced no stringent market tests.

What about investors and financiers? Presumably they had a financial stake in figuring out the best predictions, the most accurate or likely ones, and in being cautious in the face of inadequate, ignorant, or overly enthusiastic predictions. In its own technological enthusiasm, the federal government had assumed liability for most of the cost of accidents in the Price-Anderson Act, so that after 1957 there were no insurance companies with an interest in good estimates of the chance of accidents. If the costs of normal operation were a matter of guesswork, those of accidents were wild speculation, and no one had an interest in gaining better information.

Other economic calculators — the people running electric utilities, and potential investors — also lacked reliable, concrete information about the construction or operating costs of the reactors they were ordering, since there were no reactors like theirs yet completed. So they *had* to accept the manufacturers' and the AEC's predictions about the likely development of the reactors, in order to insert these into their own cost-benefit calculations. (Worse, before the oil crisis of 1973-74, regular demand growth and regulated earnings allowed haphazard economic calculations at many utilities.)

Of course it is a normal feature of development projects that there is not enough prior information to make clear judgments and calculations, so you have to rely on predictions, and these predictions are based largely on intuitions and general assumptions about how the world works. Because there is no algorithm for deciding whose predictions will be best, you tend to go with the predictions of those people you trust or like for other reasons. In the case of nuclear power, there was considerable room for both government predictions and corporate promotion to fill the gap.

Once nuclear reactors appeared commercially viable in the United States, they could be packaged for export, with the myth of commercial success as the best sales tool. In the late 1960s several industrial countries experimentally ordered American-style light water reactors. Even France, which had developed its own gas-graphite design, was taken with the prospect of a cheaper American line. The main cost evidence, of course, came from AEC promotional efforts and the increasingly pro-nuclear decisions of American utilities. Just one amusing example of how far off French expectations were: one of the cost estimates that the French used to justify ordering light water reactors was a 1967 estimate for the Diablo Canyon nuclear plant, which claimed that electric capacity would cost $146 per kilowatt (Bupp and Derian, 1978: 88). When the plant finally came on line in 1985, its electricity cost around $3,000 per kilowatt — more than twenty times as much as predicted.

Here was a technology that was inefficient as an energy source, more costly than alternatives, and extremely risky.

Although there was little evidence about costs, the Shippingport reactor went on line in 1957 producing electricity at a cost roughly ten times that of coal-fired generation (Clarfield and Wiecek, 1984: 273). Scientists at the AEC's own Brookhaven Laboratory wrote a 1958 report describing accident scenarios in which 3,000 people would die immediately, with another 40,000 injured (United States Atomic Energy Commission, 1958). Light water technology had not been developed to generate electricity cheaply or safely, but to power submarines, so that reactors had been designed primarily to be small. Yet a handful of people in government and industry managed to get this technology accepted as a major source of electricity within a decade or two. And not just in the United States, but throughout much of the industrial world. A small circle created an entire industry, and a demand for its product. They accomplished this partly through their predictions alone, and by disseminating those predictions to people who wanted to believe them. And in part they themselves went out and created a world in line with their predictions.

New Directions After the Oil Crisis

Policymakers' ability to reshape the world continued in the next stage of nuclear history, as the world responded to the oil crisis of 1973-74. When the oil crisis first struck in October 1973, France and the United States had equally ambitious nuclear plans, which the oil shock should only have confirmed. But in the months following the initial oil embargo, policymakers in the two countries reconsidered their commitments and moved in opposing directions. As in the early stages of nuclear commercialization, the predictions and policies reshaped the very factors that should have been guiding those policy choices. Policies influenced many variables that are expected instead to determine policies, such as the costs of nuclear energy, alternative energy resources, public attitudes toward nuclear energy, and the strength of the antinuclear movement. Even the policymaking structures — the very mechanisms for judging nuclear choices — themselves began to change as a result of the policies adopted.

In the same five or ten years following the oil crisis, consensus over nuclear predictions fragmented in the United States while solidifying in France. In the United States new actors entered the growing controversy, including politicians and economists (in state public utility commissions and elsewhere) skeptical of the reigning cost predictions. Ironically, the same inflated claims that had spawned the American nuclear industry were now undermining it. In France, top policymakers overrode doubts about the costs and safety of the American reactor technology, suppressed the antinuclear movement, and made a huge commitment to nuclear energy. The French managed to make their nuclear optimism a reality.

The costs of nuclear energy in the two countries were heavily influenced by government policies. The strong French commitment allowed economies of scale, standardization of reactors, lower costs of fuel processing and reprocessing, and priority in construction resources. Some of the costs were simply shifted from the national utility, Electricité de France (EdF), to the state, making nuclear construction cheaper to EdF. But others derived from the size of the commitment France made in 1973. In the United States, regulators' reluctance to intervene to improve utility management eventually led to huge cost increases. While utility management (strong in France, uneven in the United States) explains most of the cost differences between the two countries (Jasper, 1990: part 4), government policy not only interacted with that management but influenced costs directly.

As the costs of nuclear energy diverged in the late 1970s, and as they became more accurately measured, they fed back into the policymaking system. EdF convinced the French government to finance its ambitious construction program partly because costs appeared low, and low costs eventually helped make nuclear energy popular in France. In the United States, in contrast, high costs discouraged cheap financing or political support. Wildly inaccurate predictions about costs that had been made in the 1960s and early 1970s began to undermine support for nuclear energy here. The reason: those predictions were never backed up with the commitments needed to make them a reality, in contrast to France. American regulators remained optimistic and promotional, but could not spend the funds to make their dreams come true.

The influence of policymakers' expectations on the safety of nuclear plants was similar. The most important factor was again management attitudes and competence, but state policy influenced these. In France, good management was reinforced and given technical support by the close relations between EdF, nuclear manufacturers, and regulatory bodies. In the United States bad management could not be corrected by the formalistic regulatory style of the AEC and its successor, the Nuclear Regulatory Commission. Likewise the United States lacked mechanisms for screening out utilities that should probably not have had nuclear plants in the first place.

In addition to the effect of regulatory style, state support for nuclear energy may have affected management attitudes toward safety. Because nuclear energy remained controversial longer in the United States, the nuclear industry pretended for strategic reasons that nuclear energy was perfectly safe, continually underestimating the potential for serious accidents. The French nuclear industry had full government support, which allowed it to acknowledge serious risks. Thus French operations aimed to keep minor incidents from developing into major accidents and, for this reason, kept special safety teams at each plant. American managers tried to keep incidents from occurring, but always seemed surprised when they did. Given the huge difference between relatively minor incidents, which occur frequently, and major accidents, which can be catastrophes, the French approach may have produced greater public safety. EdF officials were so confident in their system that many said they almost wished a major accident would occur, so that the public would see that nuclear energy was simply a risky industry like many others rather than one which had to be absolutely perfect.

State policies also shaped the perceived alternatives to nuclear energy. The cost-benefit comparisons of energy sources instituted after the oil crisis in the United States — ranging from the econometric forecasting of the Project Independence Evaluation System established in 1974 down to the increased cost concerns of state public utility commissions — kept alternatives in the foreground of energy debates, so that the United States' energy wealth could be perceived as an alternative to nuclear. (Before the oil crisis, abundant fossil alternatives had not influenced nuclear policy.) In France less attention was paid to alternatives, in part because policymakers were optimistic about nuclear reactors and in part because EdF could expand its electric market only through nuclear expansion (nuclear-generated electricity could replace the natural gas marketed by EdF's rival, Gaz de France). The French commitment to nuclear energy has made development of alternatives unnecessary and unlikely, since excess electric capacity now exists in France.

Similarly, the demand for electricity should ultimately determine how much nuclear capacity is built, but even demand was shaped somewhat by nuclear policies. In the United States, slowing demand growth led to a curtailment of reactor orders after the oil crisis, but the disappearance of nuclear energy as a viable option further encouraged American policymakers in their efforts to reduce electric demand, especially through conservation and energy efficiency. In France pro-nuclear policies actually discouraged full consideration of techniques for encouraging conservation or demand management. There was little need. EdF was even allowed to expand demand by actively marketing electricity, for example through home heating.

State policies even affected public attitudes toward nuclear energy. In another work (Jasper 1988) I examined public opinion in France, Sweden, and the United States. I showed that attitudes in all three countries had been fairly evenly split on nuclear energy in the mid-1970s (except that American opinion was more pro-nuclear), when current policy paths were being chosen. Opinion began to diverge in the late 1970s, with each public coming to support its country's de facto policies. French public opinion grew steadily more pro-nuclear in the years after 1978, while American opinion grew more antinuclear. Sweden, where the direction of nuclear policy remained uncertain, retained a roughly even split of opinion. Public opinion was influenced by state policies more than it influenced them. In part the French public resigned itself to nuclear energy, and in part they were proud that their country seemed to handle it well (support actually seems to have risen after the Three Mile Island accident in 1979). As the antinuclear movement had demobilized by then, there were no voices on the other side. In the United States, the antinuclear movement remained active and visible. Evidence of American utilities' mismanagement of nuclear reactors only grew worse in the years after TMI. New reactor orders disappeared, and many existing ones were canceled.

Although it is often assumed that the strength and tactics of the antinuclear movements affected nuclear policies, those policies determined the fate of the antinuclear movements more than the reverse. In France the movement was unable to gain any permanent access to policymaking, and its supporters were ignored and actively suppressed until organized antinuclear sentiment all but disappeared. This outcome is predictable from the insulated structure of the French state. In the United States continued uncertainty over the fate of nuclear energy allowed the movement to survive in the form of local watchdog groups and several national organizations like the Union of Concerned Scientists. These groups found a role to play, testifying at NRC and public utility commission hearings, lobbying legislators, and monitoring existing nuclear plants. Their continued existence maintains the possibility for antinuclear views to be disseminated: antinuclear commentators were on the news after accidents such as TMI or Chernobyl, whereas in France only government spokespersons appeared, to reassure the public.

Nuclear energy policies even reshaped the regulatory and political mechanisms used to control or encourage nuclear power development. In each country a regular flow of nuclear legislation allowed new policy preferences to become embodied in the rules and institutions governing the licensing and operation of nuclear plants. In the United States a weakened commitment to nuclear energy allowed a decentralization of authority that would have hindered any potential future development, although later regulatory changes have streamlined procedures somewhat. In France procedures were simplified in order to make development easier. In general, legislation has been changed often and readily to reflect new policy choices.

Procedures that were changed in the past could always be changed again. But it might be harder, since so many factors have come to line up with each country's nuclear policy path. Each policy path, one could say, has created its own reality. It has reshaped costs, safety, alternative energy sources, public opinion, the antinuclear movement, legislation, and regulatory structures so that they support the chosen policy. Early optimistic predictions became a large nuclear industry in both countries, although a later change of heart by many American policymakers has curtailed further growth here. (Nuclear regulators in the United States remained promotional in outlook, but they are no longer allowed the free rein they once were.)

The policy paths have come to seem inevitable, since all these factors that were shaped by policy can be reinterpreted in retrospect as the *causes* of those policies. This rewriting of history is most noticeable in France, where elites now insist that they never had any choice but to embrace nuclear energy. They insist that there was never a serious political debate over the issue (on this active revision of history, see Jasper, 1992b).

Political Projects

What happened in France and the United States shows how one's own predictions can ultimately help or hurt political support for technology. French elites continued to believe in the wonders of nuclear energy more than elites elsewhere, largely because of the presence in high positions of so many people with the mathematical training and the broad engineering perspective of the Ecole Polytechnique. The French, whenever there was a decision to be made, made the decision based on extremely optimistic predictions about nuclear energy. But they followed through on their predictions with large commitments of state resources that helped make the dreams come true. And they ended up with a huge nuclear commitment that is relatively cheap, well-managed, and safe, compared to those of most other countries.

But in the different political context of United States, the same predictions (and they were literally the same predictions, the same data) eventually undermined nuclear energy. The predictions had been, of course, that costs would fall; instead they went up and up. In fact, in the 1960s and early 1970s, costs didn't rise much faster than the costs of most new technologies rise during the same phases (the exception, of course, being electronics, which do tend to have falling costs). But judged against the wild predictions of the early 1960s, nuclear energy looked as though it was not doing well. The divergence between predictions (now expectations) and outcomes made investors nervous in the 1970s, and it became a piece of evidence that the antinuclear movement could use against nuclear energy. And U.S. regulators had decided — fatally for the nuclear industry — that the predictions would come true without their active help.

In this sense, predictions came back to haunt many of the Americans who had made them.   In France, many of the predictions were just as overdrawn, but the people who had made them had enough power to make many of their predictions come true, and enough control over public discourse to keep any of the other predictions from being used against them.

Timing helps us understand the different institutional contexts in France and the United States. The American nuclear establishment created a nuclear system in the early 1960s through heroic promotional efforts. But then they backed off, thinking they had already done enough. They had not, but being so confident in their own predictions, the American nuclear leadership couldn't see the remaining problems with the light water reactor. Thus, not only was American nuclear energy damned because of exaggerated predictions of falling costs (which didn't materialize) but it was also hurt by technical flaws which were never worked out. The United States paid "the penalty of taking the lead," and France had "the benefits of backwardness," to use Thorstein Veblen's terms.

The French picked up the light water reactor about the time the American regulatory establishment dropped it, and made several of the refinements that American regulators should have required. EdF ordered most of its reactors in the mid 1970s, almost eight or ten years after the first big wave of American reactors had been ordered. The French believed many extremely optimistic predictions, but supported them with continuous institutional attention to help them come true.

Conclusion

Plans and predictions take many forms, for diverse audiences. Many predictions by economists, for example, are intended as a Cassandra-like warning, so that their authors hope very much that their predictions will be falsified by alert politicians and policymakers. I have described, in some ways, the opposite kind of prediction: a kind of hortatory prediction. In this instance, these energy planners and experts were thrilled by their predictions, excited at the possibilities, and so exaggerated them. It was a sense of their own power that made them confident they could overcome all obstacles, that made them believe in the unimaginable progress that awaited nuclear development.

Lee Clarke (forthcoming) has described another kind of government plan, based on predictions known to be false. He has examined emergency plans to be used after oil spills and nuclear accidents, dubbing them "fantasy documents" because it is clear they will not work. It is impossible to do much about an oil spill at sea once it is away from the source; there is no way to catch more than about ten percent of the oil. Large areas around nuclear plants cannot be evacuated in time to avoid radiation in the case of a serious accident. Yet there are elaborate plans for what to do in the case of spills and reactor accidents. And everyone who creates these documents, and who is responsible for carrying them through, knows that they will not work.

Predictions, like plans more generally, take many forms, often depending on their rhetorical use with particular audiences. At the least, in the case of nuclear power, we see both a continuum of evidence — from flimsy to sound — and a continuum of opinion — from the expectation that something is possible to the belief that it is impossible. In Clarke's fantasy documents there is considerable evidence that plans will not work, but there are strategic reasons to insist they will anyway. In the early days of nuclear energy's commercialization, there was not much evidence one way or the other about whether it could become competitive with fossil fuels, but the right people thought it could and would happen. In a third case, with little evidence yet the expectation that something is impossible, we might have cases such as UFOs, which few policymakers are willing to research. Only in the fourth case, with sound evidence and expected feasibility, do we have what policymaking should be: decisions that follow the evidence.

Policy rests on predictions, some of which can be transformed into plans and then into reality. Other predictions and plans have other rhetorical uses. But we should not expect that policymaking is always or even usually a simple matter of reading the evidence and following its implications. To accept this is to take away the heroism of many ventures in policymaking. It also takes away policymakers' accountability when their ventures fail.

References

Bradford, Peter. 1982. "Nuclear Hearings, Nuclear Regulation, and Public Safety: A Reflection on the NRC's Indian Point Hearings." Speech to the Environmental Defense Fund Associates, New York, 7 October.

Bupp, Irvin C. 1979. "The Nuclear Stalemate." In Robert Stobaugh and Daniel Yergin (eds). *Energy Future*. New York, NY: Random House.

Bupp, Irvin C. and Jean-Claude Derian. 1978. *Light Water: How the Nuclear Dream Dissolved*. New York, NY: Basic Books.

Clarfield, Gerard H. and William M. Wiecek. 1984. *Nuclear America: Military and Civilian Nuclear Power in the United States 1940-1980*. New York, NY: Harper & Row.

Clarke, Lee. Forthcoming. *Controlling the Uncontrollable: The Organizational Production of Fantasy Documents*. Chicago, IL: University of Chicago Press.

Hilgartner, Stephen, Richard C. Bell, and Rory O'Connor. 1982. *Nukespeak: The Selling of Nuclear Technology in America*. New York, NY: Penguin Books.

Jasper, James. 1992a. "The Politics of Abstractions: Instrumental and Moralist Rhetorics in Public Debate." Social Research. Volume 59, No.2: 315-344.

_____. 1992b. "Rational Reconstruction of Energy Choices in France." In James F. Short, Jr. and Lee Clarke, eds. *Organizations, Uncertainties, and Risk*. Boulder, CO: Westview Press.

_____. 1990. *Nuclear Politics: Energy and the State in the United States, Sweden, and France.* Princeton, NJ: Princeton University Press.

_____. 1988. "The Political Life Cycle of Technological Controversies." *Social Forces.* Volume 67, No.2: 357-77.

Lilienthal, David E. 1963. *Change, Hope and the Bomb.* Princeton: NJ Princeton University Press.

Mazuzan, George T. and J. Samuel Walker. 1984. *Controlling the Atom: The Beginnings of Nuclear Regulation,* 1946-1962. Berkeley, CA: University of California Press.

Pringle, Peter and James Spigelman. 1981. *The Nuclear Barons.* New York, NY: Avon Books.

Rees, Joseph V. 1994. *Hostages of Each Other: The Transformation of Nuclear Safety After Three Mile Island.* Chicago, IL: University of Chicago Press.

United States Atomic Energy Commission. 1958. *Theoretical Possibilities and Consequences of Major Accidents in Large Nuclear Power Plants.* Washington, D.C.: U.S. Government Printing Office.

Chapter 3

Science, Society and the State:
The Nuclear Project and the Transformation of the American Political Economy

Cecilia Martinez and John Byrne

Introduction

The first official U.S. government action regarding atomic research was the establishment of a national advisory committee by President Franklin Roosevelt in October of 1939. Roosevelt was responding to a warning contained in a letter signed by physicists Albert Einstein, Enrico Fermi and Leo Szilard, identifying the destructive possibilities obtainable from the emerging scientific discoveries related to the structure and properties of atoms.

The physicists' warning and the U.S. president's response triggered a set of events that eventually brought science and the state together as the patricians of a new era of "technological authoritarianism" (Byrne and Hoffman, 1988). Joined by the American military and the corporate sector, this partnership assumed the role of restructuring society around what were, and are, staunchly believed to be the progressive powers of science and technology. The last half of the 20th century has been shaped by and, some argue (e.g., Schell, 1982), traumatized by the Manhattan

Project. What is difficult to dispute is that this project and its institutional legacy utterly changed our world.

By most accounts, the significance of the "nuclear age" has been its revolutionary impact on international military and foreign policy. However, the transformation of political economy was no less revolutionary. In this respect, it is argued here that the Manhattan Project's institutional and organizational structure was the prototype for other scientific and technological ventures central to the Cold War experience. Far from being simply a military success, the Project was offered as a means to take American society to an even higher level of economic and technological development. The institutional transformations necessary to complete the research, development and construction of an atomic bomb set in motion a series of events which resulted in a highly centralized and hierarchical system of scientific endeavor.

Early Science Interest

Physicists Fermi and Szilard were only the most recent contributors in a cadre of international scientists that stretched back to Marie and Pierre Curie and others that were experimenting with subatomic structure and behavior throughout the late 1800s and early 1900s. Moreover, the linkage between the destructive power of a nuclear reaction and the possibility for creating usable energy was recognized early on. As Hewlett and Anderson note, the physicists reasoned from virtually the outset that "if the process [of fission] could be controlled, a new source of heat and power would be available. If it were allowed to progress unchecked, an explosive of tremendous force might be possible" (1962: 11). Given the stage of theoretical physics at the time, however, neither the physicists, and therefore Roosevelt, knew with any degree of certainty whether an atomic bomb or a nuclear power plant was feasible.

Still, the possibility that a superweapon *could* be constructed was enough to motivate the President to initiate a series of governmental actions. Roosevelt, in consultation with his military attache General "Pa" Watson, immediately created a governmental Advisory Committee on Uranium and charged it with the official duty of investigating whether and how atomic research should proceed. Initial members of the Advisory Committee included Lyman Briggs (then Director of the National Bureau of Standards) who served as its chair, Commander Gilbert C. Hoover (U.S. Navy), and Colonel Keith C. Adamson (U.S. Army).

Within the month, the Committee had completed its charge and reported back to Roosevelt that a chain reaction was a plausible, if as yet unproven, possibility. Over the next six months, both Fermi and Szilard continued to conduct their experiments, keeping both the Committee and Admiral Harold Bowen (director of the Naval Research Laboratory) abreast of their work. Supplied with this information, both Bowen and Advisory Committee chair Briggs reached the conclusion that the progress Fermi and Szilard were making was sufficient to recommend government support of a laboratory-scale investigation into the physics of a chain reaction. This recommendation propelled the first of a series of public actions over the next five years that created a government-university-corporate complex which would eventually succeed in achieving what President Truman later called, "the greatest scientific achievement in history."

The Institutional Integration of
Nuclear Weapons Research and Development

Central to the Manhattan Project and subsequent development of nuclear technology was the integration of science, the academy, the military, and industry in a common effort. While American scientific and industrial cooperation in weapons development had historical precedence dating back to the Civil War, this effort had been sporadic and largely an individual and product-specific engagement. Similarly, university relations with U.S. industry had mostly been centered on curriculum and cooperative education programs. For example, industrial concerns about the degree to which universities had been providing relevant and useful knowledge to the industrial economy had resulted in the development of new technical fields such as engineering and business (Chandler, 1977; Noble, 1977).   Yet for most of the early part of the century, industry had created and maintained its own internal research and development infrastructure independent of universities.   Indeed, industrial laboratories such as those at DuPont, Westinghouse, General Electric, AT&T, and General Motors had been the locus of industrial R&D and the envy of most university laboratories since the turn of the century.   Funding for research within the Bell system alone amounted to over $2.2 million in 1916, and with substantial annual increases this figure reached over $23 million in 1930, an amount far larger "than any single university in the country" (Maclaurin, 1949: 156, 159).

In order to proceed with the Bowen and Briggs recommendations for laboratory-scale research, it was therefore necessary to establish a new organizational basis for cross-institutional collaboration, since no single institution was capable of handling the complexities and scale of an atomic research project on its own.   The initial basis for this collaboration was provided through the establishment of the National Defense Research Council (NDRC) in May 1940.   The NDRC was created by Roosevelt through an executive order with support from scientific as well as military leaders.   The NDRC was distinctive in that it was the first time the national government would be directly engaged in defining and funding research in support of military requirements *outside* of a war context.   Among the several prominent university supporters advocating the creation of a governmental organization that could integrate the activities of universities and industry were Carnegie president Vannevar Bush, Harvard president James B. Conant, and MIT president Karl Compton.   The university group was joined by America's national science spokesman — the National Academy of Sciences president, Frank B. Jewett (also vice-president of AT&T) who similarly called for creation of a national government organization to direct the nuclear effort. The impact on research created by the new Council was both immediate and substantial: within six months, the NDRC had contracted 126 projects at 32 academic institutions and 19 corporations (Kevles, 1987: 298).

The NDRC was only a few months old when the Nazis invaded Belgium in the summer of 1940. This and other German actions provoked fears by scientists, as well as the military, that the Nazis might secure access to one of the world's largest known sources of uranium located in the Belgian Congo. Such a possibility, along with speculation about Germany's own progress on atomic bomb research, sparked another round of demands for increased national control of science in the atomic field. Two separate committees of scientists, which had been appointed by NDRC chairman Vannevar Bush to review the situation, recommended the immediate establishment of a government program in fission research. But mounting such an effort required a much larger and more certain funding source than the emergency funds that had been underwriting the NDRC at the time. A new executive order creating the Office of Scientific Research and Development (OSRD) was issued in May of 1941 in response to these concerns.

This time, Roosevelt's order established a new agency with a full-time director (former NDRC head Vannevar Bush assumed this role) and a Congressionally appropriated budget. While the mission of the OSRD was still to provide the military with advanced weapons research capability, its creation instituted a scientific research agency with relative leadership autonomy from the military on the organization and direction of research. OSRD director Bush was given the primary responsibility and authority to, as Kevles notes, "advance ideas for weapons from the germinal to the production stage" with direct reporting responsibility to the President (1987: 300). In addition, the OSRD provided for the consolidation of all atomic related projects under one agency roof. James Conant (president of Harvard University) replaced Bush as head of the NDRC, which was itself located within the OSRD, and the Committee on Uranium then became the OSRD Section on Uranium, or S-1.

Even though the physics of a chain reaction, and hence the feasibility of an atomic weapon, still remained a purely hypothetical possibility, the OSRD under Bush's direction continued to mobilize the nation's university and corporate resources. Bush undertook action to accelerate atomic research under the OSRD and to investigate the development of pilot plant operations. With respect to research, the OSRD was already simultaneously sponsoring a variety of programs, including the theoretical physics of an atomic weapon, power production, isotope separation, heavy water and graphic moderators, and nuclear fission (using uranium 235 or 238, and plutonium). In order to facilitate work on the development of a pilot plant, a Planning Board was created to specifically investigate the industrial engineering and manufacturing questions related to the production of an atomic bomb. For membership to the Planning Board, Bush recruited corporate and engineering expertise from among the largest U.S. companies, including officials from Standard Oil, the Kellogg Company, Union Carbide, and Westinghouse.

Thus, before the attack on Pearl Harbor in December 1941, the United States had already assembled an institutional structure to investigate the science and technology of atomic weapons. This structure had integrated university, industrial and military research and engineering, and had devised production and supply lines to ensure delivery of everything from uranium to flow meters and thermometers. Approximately 1,700 physicists were engaged in government-sponsored atomic research by the time the U.S. officially entered the war. A year and a half after the establishment of an advisory committee to study the status of atomic research and its implications for the production of a bomb, atomic science had become a matter of national policy.

The Manhattan Project: Science-Based
Industrial Development of a Nuclear Weapon

With the entrance of the U.S. into World War II, the atomic project acquired a new urgency. Consequently, an effort within the OSRD to pursue a more efficient and focused program of research was instituted. A week after the Pearl Harbor attack, Vannevar Bush divided the bomb project into two areas: engineering issues to be investigated by the Planning Board; and physics and chemistry questions to be researched at the universities. Six months later, in June 1942, Roosevelt authorized Bush to proceed with a full-scale effort to build the atomic bomb. With this mandate the OSRD reorganized the bomb project and authorized the newly created Manhattan District of the Army Corps of Engineers to assume management responsibilities. Brigadier General Leslie R. Groves (who had managed the building of the Pentagon) was given command of what became known as the Manhattan Project (Lauren, 1988: 60).

Before the Manhattan Project, the military-science-industry partnership had relied upon a somewhat fragmented organizational structure. University laboratories, industry representatives and high-ranking military officers had been loosely organized in "committees," "planning boards," and so on. After December 1942, however, this was no longer acceptable. With Roosevelt's approval of an "all-out effort", Groves was put in charge of a half-billion dollar budget and a "giant industrial complex" (Hewlett and Anderson, 1962: 115).

The NDRC and the OSRD had made substantial contributions in the policy and planning of a national atomic science program, but the scope of the research, engineering, and manufacturing required for an atomic bomb superseded the ability of these organizations. The Manhattan Project demonstrated a research and development scale and complexity which had not yet been experienced in the industrial world. Emphasis on organization at this stage was absolutely critical since many of the components of atomic bomb research and production were not yet understood. The Manhattan Project set out to simultaneously engage in theoretical physics research, the engineering and design of multi-facility pilot plants for the production of bomb grade material, and the manufacture and testing of the bomb itself. To successfully build the bomb, Groves would require the assistance of some of the nation's largest corporations, the cooperation of the country's most prestigious universities and the governmental authority to plan, coordinate, and administer the effort.

Of critical importance to the design of the Manhattan Project was the issue of management over industrial activities. Although the Project's objective was obviously military in nature, the question of subatomic behavior was scientific in character. Few in government or universities had the experience of building and managing large-scale production facilities needed to bring the project to fruition. Groves therefore looked to the industrial sector for the solution and ultimately settled on the DuPont Company for the job. DuPont was selected not only because of "its size and experience," but also because it had developed an internal organizational structure capable of integrating "all the complex activities of the company . . . around the manufacture of products" (Hewlett and Anderson, 1962: 187-188). Given the multiple activities of the Manhattan Project, this was considered essential to its success. In addition, the company was a veteran of military production and the War Department had already assigned to it the construction and operation of several explosive plants. DuPont acted immediately and personnel from its Explosives, Ammonia, Chemicals and Engineering Departments were assembled to supervise manufacturing and engineering operations of the Manhattan Project. On these matters the company had insisted, and received, complete management control in order to avoid "the many headaches of co-ordination and administration which plagued most joint enterprises between university research groups and industry" (Hewlett and Anderson, 1962: 188).

Finally, because centralized decision-making was considered critical, it was decided that overall decisions would be exercised by only three individuals: General Groves, MIT President Compton and the DuPont Company's Roger Williams. In this way, the Manhattan Project provided a merger of two modes of "efficient" organization: the multidivisional, horizontally and vertically integrated production structure of the modern corporation; and the command and control structure of the military system. The university's role was the supply of specialists to the new organization. Indeed, efficiency and organization would *have* to guide the project if and until the technical, military and political "ends" were resolved. The Manhattan Project became the prototype of contemporary technocratic order: a large-scale, multi-dimensional organization that embodies hierarchy *and* the flexibility of innovation-oriented R&D.

The physical and engineering accomplishments of the Manhattan Project were indeed impressive. Three major production and research facilities in Oak Ridge, Tennessee, Hanford, Washington and Los Alamos, New Mexico were constructed in three years. Entire new towns, and their accompanying infrastructure of houses, roads, rail lines, schools, commercial and retail establishments, water and power plants, as well as the experimental research pilot plants themselves, were built literally from the ground up. By 1945 the Manhattan Project employed thousands, from assembly workers to physicists and managed large complexes with an annual budget of approximately $2 billion (in 1995 dollars, this would be equal to almost $18 billion, an enormous sum for that period).

However impressive its size and sophistication, the truly remarkable accomplishments of the Project lay in its organizational triumphs. From the University of California, Berkeley to the University of Chicago and Columbia University, and from International Nickel and American Harvester to General Electric and DuPont, the Manhattan Project was able to integrate vastly disparate, autonomous and geographically distant organizations for one common purpose.

On July 16, 1945, five and a half years after the Fermi-Szilard letter had been delivered to Roosevelt, the consortium achieved its objective — the explosion of a twenty kiloton plutonium bomb. The first test was conducted in Alamagordo, New Mexico, and the blast was visible in the city of Albuquerque approximately 125 miles from the test site. The "release of nuclear energy," the principal goal of the atomic bomb R&D (Hewlett and Anderson, 1962: 377), was so violent that J. Robert Oppenheimer, the lead scientist in the project, was moved to quote from the Hindu Bhagavad-Gita: "I am become Death, the shatterer of worlds" (Kevles: 1987: 333).

In one of the politically most interesting debates of the modern era, and one which still causes public dissension to this day, the technical, corporate and military elites of the Manhattan Project indulged in a protracted discussion in the spring of 1945 over the proper use of this new weapon. Their ultimate decision was memorialized on August 6, 1945. On that day the prophecy which Oppenheimer had spoken of less than a month earlier came to a fiery rest for over 200,00 Japanese people in the city of Hiroshima. The world learned about the bomb blast from a statement delivered by President Truman (1945: 4):

> Sixteen hours ago an American plane dropped one
> bomb on Hiroshima . . . It is an atomic bomb. It is
> a harnessing of the basic power of the universe. . .
> We have spent two billion dollars on the greatest
> scientific gamble in history — and won.

On August 9, 1945, "the force from which the sun draws its power" was loosed again, this time upon the people of Nagasaki. Within a month, opposite sides of the earth had each witnessed the "unforgettable fire" of atomic energy (Nippon Hoso

Kyokai, 1977).   But their reactions, not surprisingly, were markedly different. The U.S. account was documented by William L. Laurence, the sole "official" Manhattan Project reporter designated to record the Project by General Groves (1945: 1, 16):

> It was as though the earth had opened and the skies had split. One felt as though he had been privileged to witness the Birth of the World — to be present at the moment of Creation when the Lord said: Let There be Light . . . A great cloud rose from the ground and followed the trail of the Great Sun. . . For a fleeting instant it took the form of the Statue of liberty Magnified many times.

> They clapped their hands many times as they leaped from the ground — earth-bound man symbolizing a new birth in freedom . . . The dance of the primitive man lasted but a few seconds during which an evolutionary period of 10,000 years had been telescoped.   Primitive man was metamorphosed into modern man — shaking hands, slapping each other on the back, laughing like happy children.

The Birth of the Nuclear World had indeed brought about a profound changes, the nature of which the Project cadre could not yet even imagine. But the exuberance with which most in the U.S. first greeted the arrival of modern man was not to be replicated in Japan. There, the exodus of primitive man was bid only in silence (Hiroshi, 1984: 60):

> No one wept
> no one screamed in pain
> none of the dying
> died noisily
> not even the children
> cried
> no one spoke

A New Technological Order

With the bombing of Hiroshima and Nagasaki, U.S. society had indeed realized a new Technological Order with many, both inside and outside the realm of the new technological elite, lauding the physicists for inaugurating a "Pax Atomica" (Kevles, 1987: 392). It was clear that, henceforth, any consideration of the national interest would take place within a nuclear context. The work of the Manhattan Project was not limited to the production of a bomb; rather, it had brought into being a new social reality governed by scientific and technological values. Science, industry, the state and the military had worked together to uphold and promote the value of "the one best means" (Ellul, 1964: 21), and hence had achieved their common goal.

In addition to the creation of the atomic bomb, the nation's scientific and technological resources had been mobilized to deliver a host of other technologies, from advances in biological and chemical warfare to microwave radar. The military, in particular, which prior to World War II had been skeptical about the intrusion of science and scientists into the business of war, had become convinced of their indispensability in maintaining U.S. control in the post-war era. A counsel to Army Chief of Staff Eisenhower expressed the issue succinctly in 1946 (quoted in Allison, 1985: 290):

> The lessons of the last war are clear. The military effort required for victory threw upon the Army an unprecedented range of responsibilities, many of which were effectively discharged only through the invaluable assistance supplied by our cumulative resources in the natural and social sciences and the talents and experience furnished by management and labor . . . This pattern of integration must be translated into a peacetime counterpart which will

not merely familiarize the Army with the progress made in science and industry, but draw into our planning for national security all the civilian resources which can contribute to the defense of the country.

It is important, however, to note that the reach of this apparatus was never narrowly construed as a military one. Quite the contrary, the civilianization of the Manhattan Project was embraced after the war as a model to guide science- and technology-based national economic development. Indeed, the appeal of the new Technological Order was, according to Seymour Melman, its promise of general prosperity (1974: 16):

From their experience with World War II, Americans drew the inference that the economy could produce guns and butter, that military spending could boost the economy and that war work could be used to create full employment. They observed that these results had not been achieved by the effort of President Franklin Roosevelt's civilian New Deal.

What Melman called "pentagon capitalism" would succeed, supposedly, where the New Deal had not.

In the years following the war, the influence of the atomic bomb complex would have on the American political economy was already evident. Only a few corporations and an even smaller number of universities would dominate the postwar technological and economic order, as they had during the war period. Spending on the plants and laboratories of the Manhattan Project had catapulted the federal government into a leadership role as its proportion of total science expenditures (public and private combined) increased from 18 to 83 percent. The dramatic increase in society's scientific investment had been concentrated in only a few organizations: "sixty-six percent of wartime contract dollars for research and development went to only sixty-eight corporations, some 40 percent to only ten. OSRD spent about 90 percent of its funds for principal academic contractors at only eight institutions" (Kevles, 1987: 342).

The highly concentrated and hierarchical research and development system that emerged out of the Manhattan Project raised significant questions: were these traits to be accepted as the necessary accompaniments of a postwar political economy? And relatedly, was the high degree of military participation in scientific and economic affairs an unavoidable consequence of the Nuclear Age? The answers to both of these questions became clear soon after the war.

The cooperative effort of the Manhattan Project had proven itself by the significant advantages it produced for science, industry, the military and the state: the average annual federal investment in research grew from $68 million in 1938 to $706 million in 1944; a market for parts and equipment as well as federally subsidized research and development was virtually guaranteed to corporate industry (in addition to special consideration in antitrust matters); an infrastructure necessary for the support of the military R&D system had been created; and, of course, the nation had put an end to the war in the Pacific.

Initial resolution of how to ensure the continuation and viability of the massive atomic energy and weapons research and manufacturing system was secured after brief but intense public discussion and congressional debate. Preliminary agreement on the matter was reached with the passage of the Atomic Energy Act in 1946. The enabling legislation of the Act formally established the Atomic Energy Commission (AEC) as the governing body for the nuclear program and put in place a "peacetime" administrative structure for the atomic energy industry. The Act granted the AEC sole ownership and control over the production and use of fissionable materials; authorized it to sponsor basic and applied research as well as the development of militarily necessary nuclear projects; and, finally, vested it with responsibility for the promotion and commercialization of nuclear-generated electrical power. It was also decided that the atomic energy-weapons complex must be governed by those few who were in possession of the requisite scientific expertise, under a military umbrella meant to ensure the nation's security. The Commission, composed of five members serving six-year staggered terms and representing leaders of both the public and private sectors, was to constitute the leadership of this elite system of national technology development and management.

The complete "civilianization" of the atomic energy complex was, of course, considered impossible. Yet, with atomic weapons now the basis of national and international security, operating and, and, most importantly, maintaining control, of a weapons research and manufacturing system was considered essential. Only the wartime Manhattan participants, as David Lilienthal, the first AEC chair, later explained, knew and understood the "official mystery and complexity of atomic energy. They were the experts; they knew it all; it was over the heads of the public and public critics were viewed . . . as a 'bunch of housewives'" (1980: 30).

But the significance of the AEC model of governance was, and has been, its transformation of U.S. capitalism and its elevation of scientific and technocratic interests, at times even over democratic aims and aspirations. The new institutional structure of the AEC encouraged a transformation comparable in political terms to the technical and scientific achievements of nuclear fission, namely, a modern state in which democratic processes and structures would now be expected to adapt to scientific and technocratic realities. Not only did the governance and administrative structure of the AEC blur public-private distinctions, it also legitimated government by expert, and acknowledged the permanent centrality of the military in a nuclear world.

The Atomic Energy Act accomplished this, in part, by authorizing the creation of the AEC General Advisory Committee (GAC). The Committee, which was composed of scientists with knowledge and expertise of the atomic field, was charged with "advis[ing] the Commission on scientific and technical matters relating to materials, production, and research and development" (Sylves, 1987: 16). But the GAC was not confined in its advisory role to purely technical matters. Indeed, the GAC was often called upon to consult and advise on military and national security issues, as well as private and public nuclear research and development.

But, perhaps the clearest political impact of atomic energy was the creation of the Joint Committee on Atomic Energy (JCAE) by the Congress. This body recognized as early as October 1945, that in regard to atomic matters "an extraordinary legislative device was essential" (Green and Rosenthal, 1963: 3). Because of the secrecy requirements and issues of national security surrounding atomic research and development, traditional ideas of democratic governance, based on an informed citizenry and public consent, were judged to be inappropriate for atomic decision-making. As a result, a new and innovative legislative device, the JCAE, was created by the Atomic Energy Act, and given "full jurisdiction" over all matters relating to the AEC and to the atomic program. Henceforth, only members of the JCAE, on behalf of Congress, would be privy to the details of AEC policy, and only they would have partial access to information in several areas of atomic energy and weapons research and development. It was also explicitly decided that members of the congressional military committees would *not* be allowed to exercise oversight responsibilities regarding the program. In lieu of public accountability and participation, the hallmarks of a democratic society, expert bodies integrated with limited congressional oversight had become an institutionalized in American government (Green and Rosenthal, 1963: 199):

> (T)he JCAE did not conceive its mission to be one of informing Congress, or of stimulating congressional and public discussion of atomic energy. On the contrary, the Committee's attitude seemed to be that the atomic-energy program could be debated in Congress only by those with immediate responsibility who were already privy to atomic secrets . . . The Committee took its commitment to preserve security so seriously that almost no information of substance was communicated to the rest of Congress.

Realizing the Nuclear Vision

During the 1950s through the 1970s, the entire atomic energy and weapons R&D program, insulated from public scrutiny and criticism, became a dominant model of national science technological development. Its position in energy R&D alone was such that by the "outset of the 1970s, 86 percent of all federal energy R&D policy funds that had been spent since World War II had gone to [the AEC]" (Lambright, 1976: 33). Few, if any, questioned this new "integrated" model for achieving scientific and technological progress. Universities not only appealed to the AEC for greater access to the expensive machinery in operating laboratories, but they also entered the competition for their own AEC funded high-energy research equipment. New laboratories modeled after Oak Ridge, Hanford and Los Alamos were created, including Lawrence Livermore Laboratory, Berkeley Radiation Laboratory, Argonne National Laboratory and Brookhaven National Laboratory.

The most valuable of the AEC's assets were the various laboratories and production facilities assembled for the Manhattan

Project. Their importance was recognized by political, military and scientific leaders alike as essential for U.S. technological development. At the same time, however, these leaders also suggested that a new age of science as social possibility was being forged. "(I)f progress in nuclear physics is important to the nation, to the world," Lee A. DuBridge, a GAC member, advised, then the national laboratories "are not whims of crazy scientists but are part of the necessary fabric of the atmosphere in which science flourishes" (DuBridge, 1946: 13).

By the early 1950s the AEC was devoting significant attention to the non-military applications of atomic research. Thus, in addition to research on the application of atomic weapons, AEC laboratories were busy conducting research on reactor development and basic research in high-energy physics. While the development of atomic reactors presented the Commission with a problem of high technical order, both the GAC and AEC regarded development of nuclear power as necessary and thus proceeded to develop a nuclear program based on what they believed to be "the ultimate possibility" for science in service to society (Hewlett and Duncan, 1969: 115).

In the wake of the successes of the Manhattan Project, the reality of a limitless source of energy seemed only a matter of time, given proper organization of scientific and engineering effort. The confidence which existed both within and outside of the AEC suggested that the nation's scientific and industrial elite would be equally successful in providing a nuclear generation system as it had been in building the bomb. President Eisenhower publicly inaugurated the atomic energy program in his December 1953 address to the United Nations. There, he unveiled his Atoms for Peace Program, declaring that "this greatest of destructive forces can be developed into a great boon, for the benefit of all mankind" (quoted in Clarfield and Wiecek, 1984: 184). Eisenhower was not alone in his proclamation. Even before the first prototype nuclear power plant was in operation, Alvin Weinberg, then director of the

Oak Ridge National Laboratory, was celebrating "the unborn technology" of atomic energy as the "solution to one of mankind's profoundest shortages" (1956: 299). AEC Chairman David Lilienthal echoed these sentiments when he stated, "atomic energy [is] not simply a search for new energy, but more significantly a beginning of human history in which this faith in knowledge can vitalize man's whole life" (1949: 145).

In order to achieve the goal set forward by Eisenhower, an institutional transformation as extensive as the one invoked by the Manhattan Project would be required. Among the most important elements in this transformation was the redefinition of "private enterprise" and the critical role that the state would play in maintaining and assuring the success of the private sector development of nuclear power.

Redefining the Enterprise

By the mid-1950s, leaders of several corporations had begun to challenge the publicly-led monopoly over commercial nuclear energy development. The criticism, however, was never intended to disengage the federal government from investment in nuclear energy. To the contrary, it was expected that state sponsorship would continue in this area. Instead, private industry complaints were directed at the structural difficulties they were encountering as they confronted the unique attributes of nuclear technology and the failure of the state to relieve them of these burdens.

To some extent, this debate was intertwined with both Cold War ideological arguments as well as earlier differences over the proper mix of public versus private systems of electrical generation and delivery. The latter arguments had surfaced periodically since the 1930s, most notably over the creation of the Tennessee Valley Authority and the Bonneville Power Administration. Cries of socialism in the power industry had been used to limit large-scale public power, as well as to thwart small-scale cooperative and municipal projects all around the country. But the issue of private versus public power was radically transformed with the prospective integration of nuclear power into the nation's energy system.

Before Lilienthal's term as AEC Chairman ended, he and Phillip Sporn of American Electric Power initiated the establishment of an industrial advisory committee for the purpose of making recommendations on the relevance of nuclear power to the industrial sector. Additionally, communication between those working on power reactor development within the AEC and corporate officials led to the formation of study groups aimed at developing proposals for private industry participation. Not surprisingly, the corporations involved in these study groups were veteran Manhattan Project/AEC partners. Members included the Monsanto Company, Union Electric Company, Dow Chemical and Detroit Edison. Soon after, in 1951 the AEC announced the creation of an Industrial Participation Program aimed at furthering the corporate role in nuclear power. Proposals came forward immediately, including two submitted by the study groups, two others offered by Commonwealth Edison of Chicago, and a joint venture proposed by Pacific Gas and Electric and the Bechtel Corporation.

Notwithstanding these proposals, it soon became clear that further participation by the private sector was at a stalemate. The Atomic Energy Act had been designed in a period dominated by the development of atomic weapons rather than the potential civilian use of nuclear fission. Issues of national security had preoccupied legislative discussion and, as a result, the 1946 Act essentially mandated that the atomic program be a government monopoly. By the early 1950s, revision of the Act became a priority in Congress and within the AEC. Both Congress and the AEC, as well as nuclear proponents within the private sector, agreed on the necessary thrust of the revisions: to create the institutional conditions and guarantees that would lead industry to take on a more substantive and authoritative role in the development of nuclear power. Among the immediate roadblocks to this end were "questions regarding patents, use of source material, international cooperation, use of classified information, and monopolization by those companies that had gained competence by holding AEC contracts" (Mazuzan and Walker, 1984: 25).

In 1954, a revised Atomic Energy Act was passed by Congress, paving the way for further public-private integration. The revised Act provided for private ownership of nuclear reactors and the licensing of nuclear materials, enabling at least some corporations to adopt their own initiatives for nuclear power (Green and Rosenthal, 1963: 13) The provisions strengthened the role of corporate industry in the nuclear program, and by 1963 AEC contracts and subcontracts to private vendors for materials, supplies, and equipment totaled approximately $3.4 billion. Yet, even as the atomic energy (and weapons) program became more and more linked with the national economy, the highly concentrated and hierarchical character of the complex remained essentially unchanged. Through most of the 1960s, over half of AEC expenditures were distributed to only five industrial giants (Union Carbide, General Electric, Bendix, Westinghouse and DuPont) and two academic contractors (University of California and University of Chicago) (Orlans, 1967: 13).

The 1954 Act demonstrated how far the integration of the public and private sectors in the political economy had come. Private and public leaders alike had acted on a common definition of the legislative challenge, articulating free-market arguments in criticizing state control as the problem facing the future development potential of nuclear power. At the same time, the Act enlisted the state as the principal investor in the development of the technology as well as making it responsible for fuel procurement. Both of these were unusual roles for a piece of legislation casting nuclear power's development in the language of free enterprise.

The ability to treat commercial nuclear power as an invention of enterprise, despite all evidence to the contrary, was captured in statements made by Walter L. Cisler, president of the Detroit Edison Company, during the course of legislative hearings on the revision of the Act (quoted in Mazuzan and Walker, 1984: 29):

> The question Congress must consider is whether at this critical period in the development, industry using its own funds, will be given the opportunity to perform its natural function of seeking out economic methods of utilizing this natural energy resource and making the resulting benefits available to all in a normal manner, or whether industry is to be restricted in its opportunities by a continuation of the existing law. In our minds we must proceed along natural and traditional lines.

In addition to calling for greater industrial participation in nuclear power development, all parties had agreed that a commercial nuclear power program could only be realized "if the basis of participation is made sufficiently attractive for investment of private capital" (Alfred Iddles of Babcock and Wilcox Company, quoted in Mazuzan and Walker, 1984: 29). The most important of the conditions necessary for such investment were the public underwriting of research and development and governmental protection against corporate liability and exposure in the event of a catastrophic accident.

Though a landmark in the development of the industry, the revised Act only satisfied the first of these financial pre-conditions. It did so by vesting the AEC with a unique governmental charter: the authority and responsibility to simultaneously act as both a promotional *and* regulatory agency. Promotion was, among other things, understood to mean responsibility for conducting basic research and for developing many of the components necessary for the operation of a nuclear power plant.

If the Act satisfied the research and development concerns
of private enterprise, it left unresolved the issue of liability.
Throughout the mid-1950s, a variety of study groups, commissions
and panels had been assembled to review the insurance and liability
problems posed by an emerging, but yet to be developed, industrial
technology. Francis K. McCune, general manager of the Atomic
Products Division of GE had been the first to bring the issue up
during the hearings on the revised Act of 1954 (Mazuzan and
Walker, 1984). The problem was described succinctly by McCune,
who stated that liability in atomic power "is bigger than any that
business has ever had to face." If an accident were to occur, it
was, he suggested, "entirely possible for damage to exceed the
corporate assets of any given contractor or insurance company"
(quoted in Mazuzan and Walker, 1984: 94). In June 1955, after a
more in-depth study of the problem, an Insurance Study Group
established by the AEC agreed with McCune's early assessment.
In a preliminary report they noted that the "fundamental difficulty"
of insurance for a nuclear power industry would be that "the
catastrophe potential, although remote, [is] more serious than
anything known in industry" (quoted in Mazuzan and Walker,
1984: 97).

The 1957 Price-Anderson Act was the legislative resolution
to the problem of nuclear liability. Yet, its passage was merely the
culmination of changes which had already occurred in the process
of making nuclear power an "economically viable" industry. The
structure of the insurance industry as it existed until 1955 was
incapable of providing the extent of coverage needed to adequately
address the risks of nuclear power. The amount of insurance
required could not be underwritten at the time by any single or
joint company effort. Leaders of the industry, under the guidance
of the study group, called for the organization of syndicates or
pools. In response to these recommendations, three syndicates
were formed by May of 1956: the Nuclear Energy Property
Insurance Association, the Nuclear Energy Liability Insurance

Association, and the Mutual Atomic Energy Insurance Pool. In creating these pools, the insurance industry had organized itself in a way meant to meet the unique demands of nuclear accident liability. Working together, the insurance industry was prepared to offer the nuclear industry total private property and liability insurance of approximately $65 million. "This represented an unprecedented undertaking by the insurance companies," according to Mazuzan and Walker, "[in that] the largest amount made available to other American industries had never exceeded fifteen million" (1984: 100).

Still, despite this effort by the insurance industry, and despite the lack of operating experience which, in usual circumstances was the basis for developing actuarial information, it was recognized that $65 million was hardly enough to address the risks of nuclear power plants. Utility and reactor vendors responded to the insurance industry initiatives by suggesting that because the limits of possible damage were "incalculable, nothing short of complete indemnification would be adequate if private development was to proceed expeditiously" (Mazuzan and Walker, 1984: 107). New Mexico Senator Clinton P. Anderson took the Congressional lead and began developing legislation which would accommodate the needs of partners in both the insurance and nuclear power industries.

A major obstacle in developing appropriate insurance coverage was the hypothetical nature of the process. Decisions on liability and property damage were being estimated without any experiential evidence and in advance of the development of the technology itself. Consequently, the free-market rationales offered during the debates over the Atomic Energy Act were unsuited to the proposed Price-Anderson Act. Without objection from industrial leaders, Harold P. Green, attorney for the JCAE, suggested that it was necessary to "preclude reliance upon forces of the marketplace as determinants of the rate of nuclear power growth" (Clarfield and Wiecek, 1984: 198). As was so often the case in the development of this technology, the role of the state was clearly recognized: in this case, to establish a state supported insurance and liability infrastructure sufficient to make nuclear power an attractive investment option for private capital.

Ultimately the Price-Anderson Act established $500 million in federal coverage. In addition to the $65 million provided by the private insurance industry, total coverage available was set at $560 million. Senator Anderson's staff had arrived at the $500 million figure by selecting the halfway point between zero and $1 billion since no hard evidence existed upon which to estimate a more realistic number (Mazuzan and Walker, 1984: 108). In JCAE hearings on the bill, Harold L. Price, director of the AEC's Division of Civilian Applications, discussed the agency's objective in dealing with the insurance question. The AEC, he stated, had "not approached this from the standpoint of disaster insurance to protect the public . . . We are trying to remove a roadblock that has been said to interfere with people getting into this program" (quoted in Clarfield and Wiecek, 1984: 199). As an additional incentive to industry, the Price-Anderson Act also contained a ten year statute of limitations on claims. This provision was (and is) included despite the fact that the latency period associated with exposure to radioactive material oftentimes will exceed ten years.

Big Science and Nuclear Power

With the Atomic Energy Act and the Price-Anderson Act in hand, the AEC was well on its way to achieving its goal of "privatizing" nuclear power. The AEC was equally aggressive in its efforts to reconstitute the nation's scientific endeavors through its system of national laboratories. Indeed, AEC's laboratory system had become what one writer has referred to as the home of "Big Science," where multidisciplinary teams could be assembled to address almost any scientific and technical problem (Seidel, 1986: 164). Seidel characterized the capability of the laboratories by the end of the 1950s as follows: "whether the object of study was photosynthesis, reactor materials, or nuclear propulsion, they were equipped to bring to bear a range of expertise on the question" (1986: 165). In this respect, the AEC had produced a highly flexible and easily mobilized scientific research apparatus whose role was to support the emerging national political economy.

In its 1959 report to the JCAE entitled *The Future Role of the Atomic Energy Laboratories*, the AEC identified its new responsibility as "strengthening free enterprise on the one hand, and the universities as centers of education and learning on the other." Its resources, it advised, were "held in trust for the nation as a whole," poised for deployment in those times when "national needs . . . called for out-of-the-ordinary arrangements, efforts, and ability." The atomic energy field was only one area of national interest that had been and could be served by the AEC. In fact, the Commission pledged in its report, "to make room for new projects and undertakings" (quoted in Seidel, 1986: 165).

The AEC's effort to expand its role was based on the argument "that it was not what the laboratories did, but how they did it" (Seidel, 1986: 166). Examined in this context, the organization was highly successful. By the end of the 1960s the AEC had all but shed its traditional (i.e., nuclear) mission as the basis for its funding. In doing so, the agency saw its budgets increase significantly and its client list grow to include the Department of Commerce, the National Academy of Sciences, the National Institutes of Health, and a host of corporations. It was also engaged in research on such diverse topics as desalination, civil defense and carcinogenesis. The "Big Science" model, as practiced by the AEC, was so successful that its proponents were able to suggest that it was capable of providing all the necessary ingredients for the mass production of scientific and technological development in nearly all areas of society.

Thus, in its mature form, the AEC represented a new model of industrial organization and production method. While the early rationale for its existence had been predicated on national security needs, Big Science constituted a model of scientific-industry-state cooperation that was at the very center of the nation's transformation from its traditional manufacturing base to its present technology base. Seidel has suggested that the national laboratories acted as the "factories of [a] cerebral American System of Manufacturing," (Seidel, 1986: 135) and that the nuclear power program served as the "pilot" case for the high technology era. Both were fundamental in the transformation of the national political economy. Through state sponsorship, a "scientific estate" had been assembled to collaborate with industry in the production of highly expensive scientific goods (including a system of fully equipped national laboratories) and sophisticated technologies, such as power reactors, nuclear submarines, ballistic missiles and a host of laser related inventions (Price, 1965).

The AEC model in fact served its purpose so well that it was duplicated in a number of other science-based industrial fields, including that of aeronautics. A federally-sponsored science organization for research in aeronautics dates back to 1915 in the form of the National Advisory Committee for Aeronautics (NACA). NACA conducted research on aerodynamics and missiles and had enjoyed "long, close working relationships with the military services in solving their research problems, while at the same time translating the research into civil applications" (Anderson, 1976: 17). In 1958, the agency was renamed the National Aeronautics and Space Administration (NASA), and along with NACA's research staff and facilities, assumed control of all appropriate space and aeronautics projects from the Army, Navy and Air Force. Along the lines of the AEC model, NASA was charged with operating aeronautics and space research and development facilities, integrating this research with appropriate military projects, and engaging in joint ventures with industrial contractors (Anderson, 1976: 24-28). Many of the same corporations doing business with the AEC were among the largest NASA contractors, including among others, General Electric and Westinghouse. As for its university linkages, by 1970 NASA had contributed over $32 million for university laboratories and equipment, and $50 million in university research grants.

During the height of the Cold War, U.S. R&D was largely military-sponsored, with university and chemical/engineering industry partners working in conjunction with the national laboratory system to guide national technological development. As shown in Table 1, between 1954 and 1970, for instance, no less than 90 percent of public R&D funds went to DOD, AEC and NASA in support of the Big Science-Big Industry technology model.

Table 1

Federal R&D Expenditures by Agency, 1940-1971
(millions of dollars)

Year	DOD	MED/ AEC	NACA	NSF	NIH
1940-1950	4,643	2,279	256	-	-
1951	823	242	62	0.1	-
1952	1,317	250	67	0.5	-
1953	2,454	378	79	2	53
1954	2,487	393	89	3	52
1955	2,630	385	74	8	-
1956	2,639	474	71	15	-
1957	3,371	656	76	30	125
1958	3,664	804	89	33	154
1959	4,183	877	145	54	224
1960	5,653	986	401	64	256
1961-1971	80,272	16,450	41,260	2,625	7,216

DOD: Department of Defense
MED: Manhattan Engineering District
NACA: National Advisory Committee on Aeronautics
       (Reorganized as the National Aeronautics and Space
       Administration in 1958)
NSF: National Science Foundation
NIH: National Institutes of Health

SOURCES: National Science Foundation, 1958, 1971, and 1972.

The concentration of federal funds in the hands of a few large federal laboratories and corporations was characteristic of the R&D complex as a whole. In 1967, the 100 largest contractors received 65 percent of all military contracts and the top ten received 30 percent of its sales to military contracts from 1960 to 1967. Others, such as fifth-ranked General Electric, established subsidiary divisions specifically for defense contract work. As a result, nearly 20 percent of all GE sales involved DOD, AEC or NASA during the same period (Melman, 1970: 77-78). Universities likewise found it lucrative to turn their research attention to Big Science topics, ranging from the development of weapons systems to social control techniques (Melman, 1970: 100). Indeed, defense research and production problem-solving became a highly profitable academic enterprise.

Many of the universities applied the AEC model to their own institutions and developed separate laboratories and research centers for federally funded research. As might be expected, the distribution of these resources exhibited similar concentration tendencies. A congressional study revealed that in 1964 ten universities received 38 percent of all federal funds to institutions of higher learning and that 50 universities received 75 percent. Those institutions which were not participants in the elite group found innovative ways to compete for Pentagon, AEC and NASA dollars. One such initiative was Project Themis. Instituted in 1967 with a budget of $20 million, and raised to $30 million in 1969, Project Themis was designed as an effort to incorporate smaller universities into the defense R&D circuit so they might collaboratively compete for research contracts. Subject areas included a range of science and social science topics from detection and surveillance, navigation and control, energy and power, to information systems, environmental analysis, and social and behavioral studies (Melman, 1970: 100). Universities, in the words of former Michigan State University President John Hannah, "must be regarded as bastions of our defense, as essential to the preservation of our country and our way of life as supersonic bombers, nuclear powered submarines and intercontinental ballistic missiles" (quoted in Lens, 1970: 127). In brief, parity in university R&D meant not simply that public funds would be more widely enjoyed, but that the university *system* as a whole would steadily become integrated into Eisenhower's military-industrial complex.

The influence of the Big Science model was so pervasive that even those who sought to propose a different approach to the conduct of science were forced to accept its underlying suppositions. Perhaps the most significant example of this imperative is provided by the National Science Foundation (NSF), which was proposed and implemented by former OSRD director Vannevar Bush. In his well-received report to President Roosevelt on postwar scientific research, Bush outlined his program (Bush, 1980: 31):

> [NSF] should be a focal point within the Government for a concerted program of assisting scientific research conducted outside of Government. [It] should furnish the funds needed to support basic research in the colleges and universities, should coordinate where possible research programs on matters of utmost importance to the national welfare, should formulate a national policy for the Government toward scientific information among scientists and laboratories both in this country and abroad, and should ensure that the incentives to research in industry and the universities are maintained.

Relieved of the "constant pressure to produce in a tangible way," it was suggested that NSF would underwrite the needs of basic science "to explore the unknown" (Bush, 1980: 32). In contrast to the research and development activities already existing in government and industry, NSF was to support only basic research.

Despite formal statements such as these, Bush's vision was never intended to build a science independent of military and industrial considerations. Certainly Bush did not subscribe to a view of science as a source of criticism for the emergent power of the military-industrial alliance. Indeed, Bush proposed that NSF should fund research on new weapons and should support basic research in order to strengthen industrial productivity (1980: 32 and 21). The independence he sought was on the narrow question of who should decide on project funding: NSF (i.e., scientists), politicians, military officers, or industrial officials. In the broader sense, however, NSF seldom questioned the need to collaborate with the military and industry. Instead, the debate was focused then (and many believe still is today) on the relative role of the individual scientist versus scientific organizations.

Conclusion

Less than two decades after the Manhattan Project, there was little doubt, either in the nation's leadership or the public, that the future of the United States depended upon its scientific and technological standing in the international order. As Kevles notes, science as it was constructed in the post-war era, as well as the scientists themselves, were credited with being the progenitors of a progressive technological era. It was an age, according to Kevles when (1987: 391-392):

> [S]cientists were identified not only as the makers
> of bombs and rockets but as the progenitors of jet
> planes, computers, and direct dial telephoning, of
> transistor radios, stereophonic phonographs, and
> color television; when research and development,
> in what President Clark of the University of
> California called this "age of the knowledge
> industry," was believed to generate endless
> economic expansion; when electronic and computer
> firms were assumed to follow close upon the heels

of local Ph.D. programs; when Governor Edmund
G. Brown of California reported that, on the basis
of an experiment in his state, space and defense
scientists could solve problems of smog, sewage or
waste disposal, and transportation.

Given the thrall of science, and the priority assigned to the
country's over half-century commitment to atomic weapons and
nuclear power, it is not perhaps surprising that the institutions
which gave us bombs and electricity played a central role in
postwar science-based industrialization.   Micro-electronics,
communications systems, computer technologies, laser devices,
composite materials, computer-aided design and manufacture,
robotics, radiology and many other industrial fields are directly
indebted for their existence to the efforts of the atomic energy and
weapons consortium.

One of the architects of the nuclear age expressed the
enthusiasm for science and technology that has pervaded postwar
society, and in particular, captured the imagination of the nuclear
dreamer (Weinberg, 1956: 302):

I do not think it unreasonable to propose that much
of mankind's social and political tradition will
become obsolete with the full flowering of the
Scientific Era simply because all of the traditional
doctrines were conceived in an economic and
technological era which bears little relation to the
age of abundance and   moderation   which   I
envisage . . . The bitterness which has been
assumed to be part of all political struggle —
whether intra- or international — will be mitigated
because the basic conditions of life have become
easier.

The results of the Manhattan Project and the Atoms for Peace Program have failed to coincide with the prospects envisioned by Weinberg. Instead, the technology has produced a litany of social ills, ranging from the nuclear arms race, to the disaster at Chernobyl and near-disaster at Three Mile Island, and the unresolved problems of bomb plant clean-up and civilian plant waste disposal and decommissioning. Equally important, the nuclear dream has resulted in the marginalization of democratic forms of governance. In this respect, the reinvigoration of democracy is, perhaps, the most significant challenge facing society in a post-nuclear age.

References

Allison, David K. 1985. "U.S. Navy Research and Development Since World War II." In Merritt Roe Smith (ed). *Military Enterprise and Technological Change: Perspectives on the American Experience.* Cambridge, MA: MIT Press.

Anderson, Jr., Frank W. 1976. *Orders of Magnitude: A History of NACA and NASA: 1915-1976.* Washington, D.C.: National Aeronautics and Space Administration.

Bush, Vannevar. 1980. *Science--the Endless Frontier.* New York, NY: Arno Press (originally published in 1945).

Byrne, John and Steven M. Hoffman. 1988. "Nuclear Power and Technological Authoritarianism." *Bulletin of Science, Technology and Society.* Volume 7: 658-671.

Chandler, Alfred D. 1977. *The Visible Hand: The managerial Revolution in American Business.* Cambridge, MA: Belknap Press.

Clarfield and Wiecek. 1984. *Nuclear America: Military and Civilian Nuclear Power in the United States, 1940-1980.* Philadelphia: Harper and Row.

DuBridge, Lee A. 1942. "The Role of Large Laboratories in Nuclear Research." *Bulletin of the Atomic Scientists.* 2/9 and 10 (November 1): 12, 105.

Ellul, Jacques. 1964. *The Technological Society*. New York, NY: Vintage Books.

Green, Harold P. and Alan Rosenthal. 1963. *Government of the Atom: the Integration of Powers*. New York, NY: Atherton Press.

Hewlett, Richard and Oscar E. Anderson, Jr. 1962. *The New World, 1939/1946*. University Park, PA: Pennsylvania State University Press.

Hewlett, Richard and Francis Duncan. 1969. *Atomic Shield, 1947/1952*. University Park, PA: Pennsylvania State University Press.

Hiroshi, Nakajima. 1984. "The Burden." In Marc Kaminsky (ed). *The Road From Hiroshima*. New York, NY: Simon and Schuster.

Kevles, Daniel J. 1987. *The Physicists: the History of a Scientific Community in Modern America*. Cambridge, MA: Harvard University Press.

Lambright, Henry W. 1976. *Governing Science and Technology*. New York, NY: Oxford University Press.

Lauren, William. 1988. *The General and the Bomb*. New York, NY: Dodd, Mead.

Laurence, William L. 1945. "Drama of the Atomic Bomb Found Climax in July 16 Test." *New York Times*. Section A: 1, 16 September 26).

Lens, Sidney. 1970. *The Military-Industrial Complex*. Philadelphia, PA: Pilgrim.

Lilienthal, David. 1980. *Atomic Energy: A New Start*. New York, NY: Harper and Row.

Maclaurin, William Rupert. 1949. *Invention and Innovation in the Radio Industry*. New York NY: Macmillan.

Mazuzan, George T. and J. Samuel Walker. 1984. *Controlling the Atom: The Beginnings of Nuclear Regulation 1946-1962*. Berkeley, CA: University of California Press.

Melman, Seymour. 1974. *The Permanent War Economy: American Capitalism in Decline*. New York, NY: Simon and Schuster.

National Science Foundation. 1972. *Federal Funds for Research and Development and Other Scientific Activities: Fiscal Years 1971, 1972, and 1973.* Washington, D.C.: U.S. Government Printing Office.

_____. 1971. *Federal Funds for Research and Development and Other Scientific Activities: Fiscal Years 1969, 1970, and 1971.* Washington, D.C.: U.S. Government Printing Office.

_____. 1958. *Federal Funds for Science: The Federal Research and Development Budget: Fiscal Years 1957, 1958, and 1959.* Washington, D.C.: U.S. Government Printing Office.

Nippon Hoso Kyokai (Japanese Broadcasting Company). *Unforgettable Fire.* Tokyo, Japan: First Impression.

Noble, David F. 1977. *America by Design: Science, Technology and the Rise of Corporate Capitalism.* New York, NY: Oxford University Press.

Orlans, Harold. 1967. *Contracting for Atoms.* Washington, D.C.: Brookings Institution.

Price, Don Krasher. 1965. *The Scientific Estate.* Cambridge: Belknap Press of Harvard University.

Schell, Jonathan. 1982. *The Fate of the Earth.* 1982. New York, NY: Knopf, distributed by Random House.

Seidel, Robert W. 1986. "A Home for Big Science: The Atomic Energy Commission's Laboratory System." *Historical Studies in the Physical and Biological Sciences.* 16/1:135-175.

_____. 1983. "Accelerating Science: The Postwar Transformation of the Lawrence Radiation Laboratory." *Historical Studies in the Physical Sciences.* 13/2:375-400.

Sylves, Richard T. 1987. *The Nuclear Oracles: A Political History of the General Advisory Committee of the Atomic Energy Commission, 1947-1977.* Ames, IA: Iowa State University Press.

United States Joint Committee on Atomic Energy. 1960. *The Future Role of the Atomic Laboratories.* Washington, D.C.: U.S. Government Printing Office.

Weinberg, Alvin. 1985. *The Second Nuclear Era: A New Start for Nuclear Power.* New York: Praeger.

_____. 1956. "Today's Revolution." *Bulletin of the Atomic Scientists.* Volume 12. No. 8 (October): 299-302.

PART II

The Social Consequences

Chapter 4

No One Ever Told Us:
Native Americans and the
Great Uranium Experiment

Cate Gilles

Introduction

Starting in 1946, companies such as Kerr McGee, Vanadium Corporation of America, Foote Mineral, Amex and Climax began a twenty year commitment to produce bomb-grade uranium from the federal government and fuel-grade material for civilian nuclear plants (Redhouse, 1991: 2-5). Much of this uranium came from Indian land; 13 million tons was mined from Navajo land alone. In addition, almost all of the major uranium mills in the Southwest are perched precariously close to tributaries of the Colorado River. This is the river that winds its way through the desert, has cut the Grand Canyon, and serves as an artery of lifeblood for all living things in the region. The Colorado Plateau now holds more than 70 million tons of highly radioactive tailings created by uranium milling operations. These tailings will remain radioactive for 100 million years.

There's a small museum in Grants, New Mexico dedicated to the region's experience with uranium mining. Halls of displays and a life-size shaft mine excavation inside the Grants Chamber of Commerce stand in mute and silent witness to an age whose impacts will continue long after the end of the lives of the people of this region. The mine elevator is one of the most popular attractions for school children who come from all over Arizona and New Mexico. There are several photos of Navajo Paddy Martinez and his family, the man who is given credit for discovering uranium in this area.

Nice myth, Acoma poet Simon Ortiz remarks acidly. Indian people had no use for releasing the energy of uranium, nor did they run out and find it for the white prospectors. Furthermore, tribes in the area were completely unprepared for the cash economy and an industry that would lock them into regular boom and bust economic cycles. Indeed, Indian communities were targeted for uranium mines when they had no economic development alternatives (Ortiz, 1992: 534-535):

> [The Grants museum] and the U.S. system would have us believe that it was as simple as that: it would reiterate the idea of the Indian bringing his own fate upon his head . . . No, it was not that Navajo man who discovered uranium. It was the U.S. government and economic and military interests which would make enormous profits and hold the world at frightened bay which made that discovery in a colonized territory.

The photos and displays at the Grants Museum make uranium fever in the Colorado Plateau look like a heroic adventure, a mighty victory of men over rock. But the story in the museum was paid for by major mining companies who swarmed across the region during the height of the quest, including United Nuclear, Anaconda (Owned by ARCO), Homestake Mining Company, Kerr McGee, Sohio, Western Mining, Gulf Mineral and Western Nuclear. A far truer picture of the far flung radioactive contamination of the Southwest can be gained by recognizing that the uranium industry operated within what amount to nations transformed into energy colonies, with the full support of the U.S. government. Decades after the uranium industry peaked in the Southwest, a regional cumulative impact statement on the full toll of mining, milling and bomb fallout has never been completed.

The Federal agency with oversight responsibility during the early period of mining, the Atomic Energy Commission (AEC), abrogated responsibility to do anything with the massive amount of data that was already available on radioactive hazards to miners. Conveniently the agency decided not to regulate mine conditions when they opened the mines in 1948, where as many as 5,500-6,000 men, at least half of them Native Americans, were at work (Ball, 1993: 46). The government did not even begin to implement safety and health standards in uranium mines until 20 years into its "great radiation experiment" when demand for the mineral had slowed and protecting miners was less costly.

The government's greatest fear and shame was that someone might get wind of the great hoax they were perpetrating. The hoax involved manipulating and suppressing study results indicating mining could harm both host communities and miners. In spite of repeated study results showing that exposure to radioactive toxins would harm all miners, federal government policy was to hide the evidence. Their excuse was "national security"; in federal eyes, the government's "need" for uranium outweighed the human right to health (Ball, 1993: 109-110):

> In the uranium miner's tragedy, the government
> had to keep the secret of the medical literature on
> uranium mining in Europe, Africa, and Canada, as
> well as the PHS [Public Health Service] medical
> studies. Repeated . . . health warnings and other
> reports and recommendations that would 'alarm the
> public' were kept [secret] due to the understand-

able fear that miners, if they became aware of the danger to their health, would flee the mines.

As the uranium frenzy grew to fever pitch on the Colorado Plateau and San Juan Basin, where it approached the passion of the gold rush, Native people were left completely in the dark about potential hazards, and were often bewildered by technical information, which if it was made public, no one bothered to translate. Many people living in uranium mining regions simply did not understand what radiation meant since the kinds of cancers caused by contact with radiation and its associated gases, such as radon, were unheard of. Few efforts were made to correct this situation as the dangers of exposure to radiation were an intentionally neglected part of long-term community educational efforts. Thus, community people built homes with radioactive mine tailings and livestock drank from streams and in washes that regularly hosted mine dewatering effluent. All over the Navajo reservation children regularly played on tailings piles and swam in water holes that were actually abandoned uranium mine shaft.

Even today, Navajo elders discussing new mining proposals can be baffled by nuclear language. Mining activist Mitchell Capitan, for instance, reports that "he is forced to fall back on inexact terms like 'sunlight that harms' when discussing the mining with local elders. There is no word for 'radiation' in the Navajo language" (Brenner, 1995). The dangers caused by past mining practices continue even today: mines that aren't covered continue to emit radioactivity (Gilles, Reed, and Seronde, 1990).

This chapter documents the results that the parallel policies of secrecy and ignorance have had on the Native peoples of the American Southwest. In particular, the chapter focuses on the Navajo and Havasupai Nations and Northern New Mexico's pueblo communities. It also offers a critique of the meager effort to compensate the victims of these policies, the Radiation Exposure Compensation Act.

Uranium and Native Communities of the American Southwest

The Navajo Nation

The Navajo nation of New Mexico and Arizona have perhaps suffered the most egregious harm due to uranium mining and milling in the American Southwest. The harm has affected all aspects of Navajo life: the land has been ravaged, the people have died early deaths, future generations have been imperiled, and the community has suffered by a hasty and ill-timed integration into a larger cash economy.

The contamination of both the air and the water is the most obvious legacy of uranium economics. On the northeastern edge of the Navajo nation, a Kerr-McGee mill bled toxins into the San Juan River and contaminated Farmington's water supplies. After the mill was abandoned by a later owner in 1968, "the wind blew the fine dust into Navajo houses; rain washed tailings into the river; children played on the piles; and . . . many Navajos constructed their houses with tailings material" (Ambler, 1990: 178).

The practice of using of rivers and aquifers as "waste dumps" was an integral part of uranium operations. Beginning in 1950, mine dewatering was standard operating procedure in New Mexico. Rather than treating the water to remove radioactive and other toxic elements, however, the companies simply disposed of the contaminated water by dumping it into the region's rivers and open areas, including the Rio Puerco. This river winds its way through a number of small Navajo communities that are home to more than 8,000 people. At their peak, mines along the river released more than 8.3 million gallons of contaminated water per day. During those years standards for gross alpha and beta particle activity and radium 226 exceeded safe water standards more often than not (Shuey, 1992: 4).

When a dam at the Churchrock uranium mill on the eastern side of the Navajo Nation broke in 1979, it immediately released 94 million gallons into the same river. This "hot" liquid was saturated with uranium, selenium, cadmium, lead and other heavy metals. It was the largest accidental spill of radioactive toxins in the world prior to the Chernobyl accident. Nevertheless, compared to the coverage of the Three Mile Island accident, the disaster that struck the Navajos went nearly unseen by the eyes of the society at large (Ambler, 1990: 175).

After the Churchrock spill, Navajos living along the banks of the Rio Puerco feared their own drinking water and its effects on their animals. United Nuclear's response to the accident was to post warning signs in English. As community members noted, the signs did not deter livestock from drinking. Despite the efforts of Rio Puerco communities, the company never fulfilled its promise to supply all residents with clean water. According to one leader of the local lobbying effort, while "correct documentation" is lacking and difficult to obtain, "after the spill livestock started getting sick . . . Sometimes when it is butchered the meat smells bad and it's not fit to eat at all" (Gilles, 1989: 1).

Contaminated local water supplies is not the only legacy of uranium exploitation on Navajo lands. In 1983, the EPA ruled that the thousands of mines left abandoned on Indian lands were too remote to "pose a national risk." The agency determined that mines on state lands should be reclaimed by states. That left tribes hanging, with no one responsible for cleanup of abandoned mines on reservations (Ambler, 1990: 179). Twelve years later, Bernadine Martin, Director of the Navajo Nation's abandoned mine reclamation program, says the Navajo tribe is still counting mines and searching for monies to cover them and protect community health (Gilles, 1995: 1).

A principal reason for the tribe's reclamation difficulties is the federal government's failure to establish uranium ore as a hazardous material. The reluctance to increase the number and scope of regulations on a material with substantial profit potential is easy to understand: additional regulations would increase the costs of providing for the safety of those mining uranium. The ultimate result of this regulatory practice has been to substantially reduce mine reclamation requirements on federal and tribal lands (Gilles, Watahomigie, and Bravo, 1991: 4).

The costs of uranium mining and milling are also being passed onto future generations. Arizona's statewide birth defect rate between 1969 and 1980 was one-third higher than the national average. And the Navajo communities of Cameron and Grey Mountain have experienced birth defects five times more frequently than the national average. On the eastern side of the Navajo reservation, Dr. Lora Shields and other physicians have documented upward shifts in the ratio of male to female births, cancers in non-miners and children, miscarriages and unexplained infant deaths in uranium mining areas (Shields et al, 1992; Gilles, Watahomigie, and Bravo, 1991: 5-6). All of these communities hosted numerous abandoned uranium shaft mines. When they filled, children swam in them and livestock drank the water.

A 1991 gathering of scientists looking at genetic damage in uranium mining districts also recorded measurable differences between white and non-white birth defects in New Mexico. While the white population of the state experienced 993 birth defects per 100,000 live births (compared to a U.S. rate of 841), the Native American population had a rate of 2,600 per 100,000 live births. San Juan County, which contains the Grants uranium mining district and the eastern side of the Navajo reservation, had a rate of 6,000 per 100,000 births (Wiese, 1981: 55-56). In addition, the researchers conducting this study stressed the fact that birth records only record major defects and more subtle defects go unrecorded even if they are detected. As a result, this method of recording birth defects "grossly underestimat[es] the true rate of congenital anomaly . . . probably . . . by a factor of four" (Wiese, 1981: 55).

Non-Indian communities flourished and grew in response to the booming uranium industry and shrank in response to worsening market conditions, layoffs and mine and mill closures. But those communities and miners weren't tied to the land the uranium came from. For them, moving on for jobs was a way of life. Native nations such as the Navajo live where they have always lived and therefore bear the brunt of uranium related contamination. There is nowhere else to go. Attorney Jerry Strauss remarks that the pattern of harm experienced by the Navajo nation fits neatly into the five hundred year history of colonial invasion. "It must be remembered that Indian tribes are sovereign nations defined, in large part, by the lands they occupy. Indian tribes cannot relocate their sovereign power to new territory if their existing territory is made uninhabitable. If we destroy tribal territory, we strike at the very roots of tribal existence" (quoted in Eichstaedt, 1994: xi).

A new proposal to leach uranium from the groundwater is continuing this tradition of exploitation. Texas-based Hydro Resources Inc. would like to leach uranium from the water below several sites including the aquifer beneath the site of the 1979 Churchrock spill. Company spokespeople are adamant the process is safe.

The Havasupai Nation

Uranium mining and milling operations have regularly desecrated or destroyed places that are sacred to different Southwest nations. One such nation is the Havasupai, who make their home in the bottom of the Grand Canyon. The Canyon's rims have hosted thousands of uranium mining claims and dozens of shaft mines. In 1986 tribal members discovered that Energy Fuels Nuclear was proposing an underground mine operation on the South Rim. The Canyon Mine project is located perilously close to land that is most sacred and central to the ancient Havasupai religion.

For nearly ten years the tribe has struggled to protect its land and people through interventions in the environmental impact assessment process and by court actions, all to no avail. The tribe has never before been forced to be this public about private spiritual and cultural practices that root them in the Grand Canyon. Former Chairman Delmer Uqualla says that Energy Fuels Nuclear is interfering with long-practiced spiritual practices of the Havasupai people. "We were promised that we would be able to continue to practice our religion on our ancient land. If our religion is destroyed, how can we practice? . . . We were promised access to our sacred sites. Right now EFN has a fence around one of them and we can only get in with permission from the mining company" (Gilles et al, 1991: 7).

A few days before Earth Day 1990, a federal court judge ruled against the tribe's religious freedom and environmental protection claims. By finding for the mining company, the judge effectively reaffirmed the dominant society's valuation of private property rights as more important than ancient Native American religious traditions. The Ninth Circuit Court affirmed the lower court ruling and dismissed the tribe's appeal. The Supreme Court refused to reopen the case, leaving the Havasupai with no legal option to protect land that for them is some of the most sacred in the world. In 1991, the tribe adopted a new constitution that permanently bans uranium mining activity within reservation boundaries. Unfortunately this document has no power over the Canyon Mine or any of the mines on the North Rim, since all of them are located outside reservation boundaries.

The Tribe has also challenged the Forest Service's acceptance of company assertions that mining poses no threat to the aquifer that supplies the only pristine water the tribe has access to. Many of EFN's safety claims are based on the idea that there will only be one mine operating south of the Grand Canyon.

However, if the industry realizes a significant upturn in demand, there could be dozens of mines operating on the South Rim alone (Gilles, 1991).

New Mexico Pueblos and the Grants Uranium Belt

The Grants Uranium Belt of northwestern Mew Mexico is home to thousand-year old Pueblo communities. During the 1970s, the mine memorialized in the area's regional museum was the largest uranium producer in the United States. Mining and milling brought abrupt and radical change to many pueblo communities located in the district. Those changes were largely negative: rising suicide rates, spousal abuse, alcoholism, and sexual abuse of children in staggering numbers. According to Aacqumeh Hanoh (Acoma) activist and poet Ortiz, who watched uranium infiltrate his own community (1992: 356):

> The Laguna miners would find themselves questioning how much real value the mining operation had when their land was overturned into a gray pit miles and miles in breadth. They would ask if the wages they earned, causing wage income dependency and the royalties received by the Kawaikah people were worth it when Mericano values beset their children and would threaten the heritage they had struggled to keep for so long.

The first deal that the Laguna people made with Anaconda was so casual that the paperwork was kept in a suitcase. "Although the Laguna Pueblo held a vote on the proposed mining, few of the tribal members could read the referendum question, much less comprehend the impact on the land from uranium development" (Ambler, 1990: 58-9). Since that time, researchers have documented massive social shifts as the pueblo's agricultural economy was rapidly transformed by the cash-based mining economy. Acoma scholar Manuel Pino argues that the mining accelerated the shift from traditions to consumerism, from the Laguna language to English (Eichstaedt, 1994: 163-5). According to Pino, the illnesses and social problems that resulted from mining in the Grants Belt have never been addressed. While numerous studies record the fact that too-rapid resource development on Native lands is intimately linked with community destruction (LaDuke, 1992; Shkilnyk, 1985; Dawson, 1992), Pino argues that (Gilles, 1994: 1):

> Very little research looks at the destructive socio-cultural process that uranium mining has imposed on Indian people. We believe that when you destroy the land, you destroy the people. And in many cases this has destroyed us, so you can understand why we don't trust this government or the agencies that are supposed to regulate the nuclear industry as a whole.

The Radiation Exposure Compensation Act

The eventual contamination the Grants Belt people faced when the Anaconda mine stopped operating was mind boggling: they had hosted the largest uranium strip mine on the planet for nearly thirty years (Ambler, 1990: 182). The mining operations created three open pit mines, thirty-two waste dumps, twenty three subgrade ore stockpiles, and nine underground mines. During 29 years of mining, over 400 million tons of earth were moved. After lengthy negotiations between the tribe and Anaconda, the mine, which eventually consumed 2,656 acres of the pueblo, is now in the final stages of reclamation. Reclamation does not, however, include plans to study the human health effects experienced by Native miners or their communities. Nor are any agencies planning long-term monitoring at the mine, which spreads across the land near the pueblo.

The limitations of "recovery" are explained, in part, by the deficiencies of the Radiation Exposure Compensation Act (or RECA). To some extent, the Act represents a formal apology by the United State Government to those individuals and communities who were sacrificed in the name of national security. Critics find two major problems with the Act. First, many of the victims of federal uranium production policy will never be acknowledged, in that the Act excludes a great number of the people most directly affected by uranium activities, i.e, uranium strip miners, uranium millers, people living upwind of the Nevada Test Site, and children born with horrific defects in uranium mining areas. Instead, the Act compensates only people who were underground uranium miners and downwinders.

Second, the Act establishes significant hurdles both in terms of causation and documentation. As Jon Erickson remarks (1993: 6):

> Proving causation is still not an easy task, and requires health, work and marriage records many of which are not available for Native American miners and their families. According to the Tribal Office of Navajo Uranium Claims, only 54 of the 328 approved claims so far are from Navajos, and Navajos are again facing bureaucratic and legal difficulties in filing claims.

One of the roadblocks Navajos face when they apply for compensation is spotty Indian Health Service records that are hard to access and sometimes grossly inadequate. The debate over the U.S. Justice Department's refusal to accept traditional Navajo marriages illustrates the problem. By 1993 Navajo President Peterson Zah was concerned enough about the issue to include it in his state of the union address (Zah, 1993: 5):

Whether the marriage is by custom or under tribal law, we need the authorization for our courts to validate them. This amendment to the Navajo Tribal Code would especially help those who have applied for compensation under the Radiation Exposure Compensation Act. We have many families who have filed claims for compensation but lack government documentation to prove marriage. Because of this, widows are being denied compensation by the U.S. Department of Justice.

Former U.S. Secretary of the Interior Stewart Udall, who was an early proponent of compensation and had participated in a series of miners lawsuits which eventuated in the Act, charged racism in 1992 when he discovered that white miners were being compensated twice as quickly as Native miners. Udall blasted the Department of Justice (Eichstaedt, 1994: 153-154):

You have grossly neglected the applications of elderly uranium miner widows who should have been accorded places at the head of the line. The pattern of your payments reveals an anti-Indian bias . . . the regulations your staff devised concerning documentation of medical facts severely penalize Navajo applicants.

During one of the first series of health tests, only five of the 516 Navajo uranium miners who had been tested were eligible according to the federal eligibility requirements. Dr. Louise Abel, who conducted many of the tests, called many of the requirements senseless and argued the test requirements were urgently in need of revision (quoted in Eichstaedt, 1994: 240):

Some of the problems could not be addressed because they reflected inconsistencies in the law itself. Our Navajo miners worked all over the

Southwest. Many of these are having difficulty with compensation because they cannot prove enough exposure solely on the reservation . . . This does not make medical sense. Another problem that has plagued us is the regulation requiring medical documents to be certified. The IHS (a federal entity) has had difficulty meeting this requirement due to lack of staff. This has put the Navajos who rely on the IHS at a disadvantage. The Federal Government should accept its own documents without the certification requirement.

Limiting compensation to only a few of the injured parties and excessive and unworkable administrative demands are only a part of the problems with the Act. Perhaps even more important is the Act's failure to deal with the more extensive *social* harms which resulted from the mining and milling activities. Marjane Ambler's research, documented in *Breaking the Iron Bonds,* has demonstrated multiple levels of uranium-related destruction and the inherent inadequacy of the Act (Ambler, 1990: 183):

The federal government typically weighed lives against dollars and cents and set its standards by determining how many deaths would be considered 'acceptable.' Unlike the statisticians in Washington the local people could see the costs of the continual exposure to low-level radiation. Lung cancer reached epidemic proportions among uranium miners on the Navajo reservation and studies showed clusters of birth defects around the areas with the most abandoned uranium. A person who died of lung cancer was listed as a "negative health effect" in government studies, but to the Navajos he was a neighbor or a clan brother.

For many miners, the colonial legacy is the fear they feel every time they cough or see the doctor: many wonder if an X-ray will reveal a dark spot on their lungs. The reflections of one Native miner, Raymond Joe, illustrates the concerns experienced by whole communities. Many of Raymond's relatives died after working in the mines. Yet, "when it first started, people thought it was a good thing, that they could make money. So they said go for it. But after 20 years of the mining, people started to die," Joe says, adding: "I worked in the mines for 11 years and no one ever told us it was dangerous" (Gilles, 1990: 6).

Raymond worked in the Cove area on the eastern side of the Navajo reservation, an area pierced with hundreds of abandoned mine shafts. In the mines, his and the other worker's lives were laced and woven with hostile toxins. They were not told the awesome concentrations of radon gas in the mines should be vented. They were never told to stay away from the dust and the water trickling down the slimy mine walls. Some ate dust as spice for lunch and washed it down with radioactive water from the mine. They were never told to take their dusty clothes off away from home, never told to avoid hugging wives and children when covered with dust, never told this job could cost their lives (Dawson, 1992: 392).

After studying Navajo uranium miners and mill workers and the effects all of the radioactive secrets had on their lives, Susan Dawson has concluded that the government violated the basic human rights of Navajo uranium workers. "The Navajo worked in the nuclear industry for the defense of the United States, working in good faith and believing their lives were not endangered. The government failed in its responsibility, then to ensure the safety of their work . . . Overall, the government failed to meet its legal and moral obligations to these uranium workers" (Dawson, 1992: 396).

Radiation has also fractured the biological well-being of the circle of life in the Southwest, affecting plants, wildlife, livestock and humans. Yet, these sorts of hazards receive little if any mention in the exhibits at Grants museum. Nor does the museum provide information on the damages caused by radiation's contact with future generations. In this respect, the museum faithfully reproduces the failures of the Radiation Exposure Compensation Act.

The government continues to insist that further study of radiation exposure standards is needed. Many Native Americans, however, having lived with the destruction caused by the radiation experiments, argue that the time for study is long past. According to Navajo activist Esther Yazzie, for instance, "[T]he whole life cycle of the Navajo people is disrupted, there's widespread contamination of our water, contamination of our grazing areas. We are all human beings, we are all involved when some of us are contaminated. We are the five fingered beings, but now our babies are being born with more than five fingers and less than five fingers" (Yazzie, 1994).

Conclusion

The deserts, forests and mountains of this land are thousand year carvings by water, wind and sun. These essential places are etched through the lives and hearts of the Native people; source, sanctuary and nurturance for medicine, spirit, life itself. The deposits of uranium below the surface are even older. Some evenings even now the night sky is as indigo dense and breathless as it must have been for thousands of years. Stars pierce through as though they were brand new. But the wind brings strange dust. At several places on and near southwest reservations, coal strip mines operate around the clock. Sometimes topsoil ends up hundreds of miles away.

At the Grand Canyon visibility is down. Some of the dirt in the air comes from California cars. More of it comes from burning coal in regional power plants to send energy back to the megacities of Phoenix, Los Angeles and Las Vegas. Ironically tribes use very little of the energy their coal and uranium resources produce. Four villages in Hopi refuse to allow power lines to cross their boundaries. At least ten thousand Navajo homes in remote areas of the reservation have no electricity (Cole and Skerrett, 1995: 47-52).

At the New Mexico Mining Museum in the Grants Chamber of Commerce someone forgot to tell the story of what is still happening invisibly, now that the uranium's energy has been released, staining the future like indelible ink: marking the genes of future children and the soil and the scrawny, hardy plant life that holds the soil down. Uranium mining has also left permanent stains on relations between Native and non-Natives here. Nowhere in the museum is there a voice asking what happens next, or asking who pays the price for societies greed? Visitors can, however, activate tapes that explain the many benefits of uranium, and the "fact" that uranium mines closed due to (irrational) public fears about health hazards.

By the time mining companies arrived in this century, Southwest nations were already damaged by decades of war; ever-increasing restrictions on their lives, spirituality, and land. The values the mining companies brought with them often functioned like bombs in slow motion, blowing up communities, separating people from values and tearing them from what had been living and sacred relationships with their environments. The boom and bust cycles of uranium profits expanded the damage. When the abrupt shift to cash had penetrated tribes, many were left high and dry, without many options when the cash vanished.

In the early years of uranium development, tribes found that the federal agencies assigned the responsibility of protecting their environment stubbornly refused to enforce laws. "Too often tribes found that the Bureau of Indian Affairs, the Department of Energy, the Environmental Protection Agency or state environmental agencies let energy industries break federal laws without rebuke" (Ambler,1990: 174).

Now, tribes are beginning to assume control over mineral leases and royalties and environmental regulation. The Navajo led the nation in pressuring Congress to implement mill tailings cleanup standards (Ambler, 1990). Yet, most discussion of abandoned mine regulation on public land remains vague. "There is no firm definition of 'acceptable' standards for reclamation, rehabilitation, containment, and stabilization because there is no legally enforceable assessment of the effects of uranium mining and milling on the environment or on the quality of human life or health" (Clemmer, 1984: 97). Nor is there any plan to clean up uranium mines in the United States (Eichstaedt, 1994: 148-149).

Uranium mining has cost Native nations a great deal of cultural integrity as well. Time and again indigenous science, knowledge, and sacred land based traditions are dismissed. Settler logic argues the only valid discourse must be rendered in bureaucratese and the arcane and dissociated language propagated by western nuclear science (Hilgartner, Bell, and O'Connor, 1982). As Dene scholar Martha Johnson points out, the interaction between indigenous and Western science is often polarized (1992: 76-77):

> The main difference between the two systems appears to be in the different types of information gathered, how it is interpreted and expressed, and their different approaches to resource management . . . [There is also a] tendency for both knowledge systems to be judged according to a rigid set of generalizations and a static image of the past.

Colonialism, dependency and depression are reinforced by the knowledge that the government has lied, and has held the peoples lands and lives in great contempt (LaDuke, 1992; Jaimes, 1992; Shkilnyk, 1985).

One of the most frightening and frequently repeated comments in the uranium production debate is "we don't know." We don't know how to reclaim, how to establish realistic exposure, we don't know how much exposure damages plant or baby genetic integrity. We don't know. Looking at the uranium production industry in this light should make us all wonder who agreed it was reasonable to experiment with our lives. The history of the Southwest uranium industry exposes many hairline fractures between expressed values and practice (Shkilnyk, 1985: 231):

> Like a crystal that reveals its hidden structure only when it is shattered, a human community discloses its structure when the glue that held it together dissolves under the impact of events and pressures, both internal and external. It reveals the extent of its disintegration when, among other things, its own members no longer care about whether or not they produce healthy offspring and when they abdicate their collective responsibilities for the physical, emotional and spiritual survival of succeeding generations.

Southwest uranium mining and milling has an uncertain future. The final decision on whether or not to subsidize the unstable industry rests in hands far to the east, in minds and hearts that may never have seen the people here. Government nuclear regulatory agencies know their capacities for clean-up are seriously limited and that many of the scars left by mining are permanent (U.S. Department of Energy, 1995:9):

> We have large amounts of radioactive materials that will be hazardous for thousands of years: we lack effective technologies and solutions for

resolving many of these environmental and safety problems: we do not fully understand the potential health effects of prolonged exposure to materials that are both radioactive and chemically toxic: and we must clear major institutional hurdles in the transition from nuclear weapons production to environmental cleanup.

The story of destroyed lands, tainted water and twisted genetic maps of life is one the Grants museum will probably never tell. Nor will there be a room or a library on what happens to sovereignty when the land and the people are poisoned.

References

Ambler, Marjane.1990. *Breaking The Iron Bonds: Indian Control of Energy Development.* Lawrence, KS: University Press of Kansas.

Ball, Howard. 1993. *Cancer Factories: America's Tragic Quest for Uranium Self-Sufficiency.* Westport, CT: Greenwood Press.

Brenner, Malcolm. 1995. "Hearings sought on uranium plans." *Navajo Nation Messenger.* February 22: 1.

Clemmer, Richard O. 1984. "Effects of the Energy Economy on Pueblo Peoples." *Native Americans and Energy Development II.* Boston, MA: Anthropology Resource Center.

Cole, Nancy and P.J. Skerrett. 1995. *Renewables are Ready: People Creating Renewable Energy Solutions.* White River Junction, VT: Chelsea Green Publishing.

Dawson, Susan E. 1992. "Navajo Uranium Workers and the Effects of Occupational Illnesses: A Case Study." *Human Organization.* Volume 51, No.4: 389-397.

Eichstaedt, Peter H. 1994. *If You Poison Us: Uranium and Native Americans.* Santa Fe, NM: Red Crane Books.

Erickson, Jon D. and Duane Chapman. 1993. "Sovereignty for Sale: Nuclear Waste in Indian Country." *Akwe:kon Journal.* Volume X, No.10: 3-10.

Gilles, Cate. 1995. "Uranium mine draws outrage." *Navajo-Hopi Observer.* (March 1): 5.

_____. 1994. "Victim Compensation Act fails to fulfill promise." *Navajo-Hopi Observer.* (June 15):1-5.

_____. 1991. "Japanese may mine in the Havasupai area." *Navajo-Hopi Observer.* (June 12): 1-7.

_____. 1990. "Annual forum held on lingering effects of world-wide uranium mining." *Navajo-Hopi Observer.* (October 31): 6.

_____. 1989. "Havasupai battle against uranium mine to preserve their land." *Navajo-Hopi Observer.* (August 2): 6.

Gilles, Cate, Marti Reed, and Jacques Seronde. 1990. "Our Uranium Legacy." *Northern Arizona Environmental Newsletter.* Volume 2, No.3.

Gilles, Cate, Don Watahomigie, Lena Bravo. 1991. "Uranium Mining at the Grand Canyon: What Costs to Water, Air, and Indigenous People?" *The Workbook.* Volume 16, No.1.

Hilgartner, Stephen, Richard C. Bell, and Rory O'Connor. 1983. *Nukespeak: The Selling of Nuclear Technology in America.* Harmondsworth, Middlesex, England: Penguin Books.

Jaimes, Annette, ed. 1992. *The State of Native America: Genocide, Colonization and Resistance.* Boston, MA: South End Press.

Johnson, Martha. 1992. "Documenting Dene Traditional Environmental Knowledge." *Akwe:kon Journal.* Volume IX, No. 2: 72-79.

LaDuke, Winona. 1992. "Indigenous Environmental Perspectives: a North American Primer." *Akwe:kon Journal.* Volume .IX, No.2: 52-72.

Ortiz, Simon. 1992. *Woven Stone.* Tucson, AZ: University of Arizona Press.

Redhouse, John. *1991. An Overview of Uranium and Nuclear Development on Indian lands in the Southwest.* Albuquerque, NM: Redhouse-Wright Productions.

Shields, L.M., W.H. Wiese, B.J. Skipper, B. Charley, and L. Benally. 1992. "Navajo Birth Outcomes in the Shiprock Uranium Mining Area." *Health Physics.* Volume 63, No. 5 (November): 542-551.

Shkilnyk, Anastasia M. 1985. *A Poison Stronger than Love: the Destruction of an Ojibwa Community.* New Haven, CT: Yale University Press.

Shuey, Chris. 1992. "Contaminant Loading on the Puerco River: A Historical Overview." Paper presented at the Puerco River Symposium. Fort Defiance, Arizona.

U.S. Department of Energy.1995. *Closing the Circle on the Splitting of the Atom: The Environmental Legacy of Nuclear Weapons Production in the United States and What the Department of Energy is Doing About It.* Washington, D.C.: U.S. Department of Energy.

Wiese, Lawrence (ed). "Birth Effects in the Four Corners Area." Transcript of meeting held February 27, 1981. University of New Mexico.

Yazzie, Esther. July 2, 1994. Personal Interview.

Zah, Peterson. 1993. "President's State of the Union Address." Transcript reprinted in *Navajo Times.* April 22: 5.

With many, many thanks to Stan Bindell, Roberta Blackgoat, Lena, Phillip, Bonnie and Clay Bravo, Faye Brown, Joyce Bugola, Nilak Butler, C.B., Jennifer Denetdale, Corey Dubin, Reggie Deer, Joely DeLaTorre, Beth Gilles, Karen Goodpasture, Eva Grabarek, Phoebe Hirsch-Dubin, Steve Hoffman, Constance Holland, Kathleen Hope, Secody Hubbard, Amos Johnson, Gerald Johnson, Winona LaDuke, Robert Lyttle, Marcus Lopez, Anita McIlmore, Meike Mittelstaedt, Hiroshi Miyata, Karen Northcott, Simon Ortiz, Manuel Pino, Larry Preston, John Redhouse, Anna Rondon, Bill Rosse, Marley Shebala, Jennifer Siyuja, Mary Sojourner, David

Solnit, Rebecca Solnit, Mary Ann Steger, Zensuke Suzuki, Reiko Tatsumi, Aya Tatsumi, Rex Tilousi, Carleta Tilousi, James Uqualla, Don Watahomigie, Jane Williams, Esther Yazzie, and Elaine Zah.

Chapter 5

Safety, Accidents, and Public Acceptance

Phillip A. Greenberg

In the arena of U.S. energy policy, no energy source continues to be as controversial as nuclear power. Beginning in the late 1970s, the industry slowed to a standstill as utilities, capital markets, state regulatory agencies, and the American public grew increasingly wary of new nuclear power plants. Although the U.S. program is the world's largest (109 plants licensed to operate as of late 1995), only one additional plant remains under construction and no firm order for a new plant has been placed since 1973.

Most observers agree that the future of nuclear power in the United States has for some time hinged on finding solutions to the industry's four key problems: safety, wastes, economics, and public acceptance.[1] The importance of the latter has been strongly emphasized (DOE, 1987: 182, 190-95; GAO, 1989: 4). Without the support of the public, the future of nuclear energy remains cloudy at best. Yet the American people may not necessarily have the last word. In recent years, public access to the regulatory process has been sharply curtailed, particularly with regard to the

---

[1] For one example, see the comments of Ivan Selin, Chairman of the U.S. Nuclear Regulatory Commission, on May 16, 1994 to a nuclear industry conference, quoted in *Nuclear News*, July 1994, p. 26.

evaluation of the safety of plants that may be proposed in the future.

This analysis focuses on the questions of nuclear power safety and accidents and their impact on public trust. The two matters — safety and trust — are closely linked. Opinion polls have indicated that the American people's views about reactor safety constitute an important factor — some observers believe the single most influential factor — in their attitude regarding the acceptability of nuclear power (OTA, 1984: 218; 65-66; Nealey, 1990: 33 and references therein). This was true even prior to the 1979 accident at the Three Mile Island (TMI) nuclear power plant in Pennsylvania (Farhar et al, 1980: 150).

In plain language, the public does not trust nuclear power. This has not always been the case. Public opinion generally favored nuclear power through the early 1970s (Balogh, 1991: 236-238). However, by the mid-1970s support had softened (Freudenberg and Rosa, 1984), with the accidents at TMI and Chernobyl often being cited as milestones in the shift of public opinion (OTA, 1984: Ch. 8; van der Plight, 1992). After TMI, opponents outnumbered supporters for the first time in several surveys and a trend of rising opposition based largely on safety and environmental concerns has continued (van der Plight, 1992: 2-5). Since the early 1980s, public opinion polls have shown broad opposition to the construction of additional reactors in the United States (OTA, 1984: 211-212), and since 1988 the margin of opposition has been roughly 2 to 1 (Rosa and Dunlap, 1994).

The questions of safety and public trust cannot be understood in isolation. They have evolved from the unusual history of the development of the nuclear power industry and its regulation. Key events and trends were often enmeshed in broader social, political, and geopolitical landscapes that shaped and defined them. Recognizing the impacts of these larger cultural currents is essential to understanding how and why the safety controversies about nuclear power developed as they did and how all these factors have combined to shape public opinion.

In the end, the public asks a straightforward question: how safe are nuclear power plants? Efforts to provide an answer lead inevitably to other questions: What could happen in the event of a severe accident? How likely is such an accident? How much risk is acceptable? These questions turn out to be quite complicated. The public's deceptively simple question has proven extraordinarily resistant to a compelling and persuasive answer.

Some seek to answer the public's question by relying on scientific and technical analysis to demonstrate that the risk of accident appears to be low. That route is appropriate to a point. But nuclear power proponents often dismiss as ignorant or irrational those who disagree with or are unresponsive to "the facts." By so doing, proponents overlook the long history of safety controversies and downplayed hazards that have dogged the nuclear industry and molded public opinion. They also miss important differences between the public and the "experts" in the ways in which risks are perceived and assessed.

Another view — the one taken here — is that there are sound reasons for the public's discomfort, and that technical analyses never constitute a final answer but rather provide only one basis from which to begin the discussion. Scientists disagree about nuclear power and their disagreements should be illuminated via public debate. The question of whether or not to move forward with nuclear power (or with any technology that carries with it high costs and potentially catastrophic long-term risks) must be examined carefully for all its implications to society. The final decisions must inevitably be made in the political and social arena. When all is said and done, in a disagreement between experts and a reluctant public, it is the experts who must give ground, not the other way around. That is how democracy is supposed to work.

Yet one of the central problems of nuclear power is that its history has been marked by the pervasiveness of bureaucratic, federal decision-making at the expense of democratic process. That the U.S. regulatory system has been more open than most to intervenor and public participation does not mean that it has in fact been very open at all. Critics and opponents have found the system stacked against them at most turns, and have been forced to appeal for relief to the courts (with only mixed success), the media, and the public. Since the early to mid-1980s, there has been a disturbing trend to close the regulatory system even more tightly, and to raise more and larger obstacles to public participation on all matters, including issues of accidents and safety. This could cause a public that has long been mistrustful of the integrity and openness of the process to have even greater doubts about the safety and desirability of the nuclear option.

Nuclear Accidents: A Brief Introduction

The U.S. Nuclear Regulatory Commission (NRC) and the industry generally separate unplanned events at nuclear power plants into two broad categories: "incidents" and "accidents." The former are unforeseen events and equipment failures that occur during normal plant operation and that result in neither significant offsite releases of radioactive materials nor severe, permanent damage to plant equipment of the kind that could pose the threat of offsite releases. While many incidents are trivial, many others have genuine safety significance.

The NRC has developed a classification system to grade incidents according to the seriousness of their safety implications. The NRC tracks incidents by requiring utilities to submit a "Licensee Event Report" (LER) when plant systems fail to operate as designed or there is a procedural lapse or breakdown. In 1990-1992, more than 6,600 LERs were filed with the NRC. A more serious class of incident is the "Significant Event," which involves reactor equipment or operational problems that have greater potential safety implications.   These include some safety system malfunctions, and immediate, unplanned reactor shutdowns ("scrams") because of equipment failures or operator error (i.e., only some of the 537 scrams in the 1990-1992 period were classified as Significant Events).   During this three-year period, a total of 107 Significant Events occurred at U.S. reactors (Riccio and Freedman, 1993a, data compiled from NRC published sources).   Since the late 1980s, operational improvements initiated after the TMI accident resulted in declining industry average numbers of Significant Events that leveled out in the early 1990s (Rees, 1994: 183-186).   However, individual "problem" plants may experience more than the average.

"Accidents" are the much less probable events that seriously threaten the operation of essential safety systems or are so destructive to plant equipment that they pose a threat of major offsite releases and/or serious damage to the power plant itself. The most severe accidents are those that could disrupt the flow of vital coolant to the reactor core and result in fuel damage or melting.   An attenuated core-melt is precisely what happened at TMI.

The Three Mile Island accident is the worst that has occurred in the U.S. commercial nuclear power industry, and perhaps one of only a very few "accidents" according to those who favor a more restrictive definition.   However, there have been numerous events — never mind whether they are labeled incidents or accidents — that have had grave safety implications and, if things had gone somewhat differently, could have resulted in a severe reactor accident with potentially serious offsite consequences. In 1975, for example, a worker using a candle to look for air leaks ignited a fire at the then two-unit Brown's Ferry nuclear power plant in Alabama.   It took plant workers several hours to bring one of the reactors under control and shut it down safely (UCS, 1977: 188-191).   An equally serious safety-related accident occurred at the Salem reactor in New Jersey in 1983.

Here, the reactor failed to shut down in response to a safety control system activation; it failed a second time three days later. Regulations governing this type of incident, known as an "Anticipated Transient Without Scram" (ATWS), had been opposed by industry on the grounds that the probability was so small it wasn't worth design attention (Marshall, 1983).

In addition, over the years numerous safety-related conditions or occurrences have been revealed which in some circumstances could have lead — or in some cases that are unresolved may yet lead — to a serious accident. Examples of such problems include the GE Mark I reactor's high probability of containment failure in the event of a core-melt accident (Asselstine, 1987: 247, citing NRC sources); fire barrier material used in 79 plants that fails to meet performance specifications by a wide margin (Nicodemus, 1994); and gauges used in 34 plants to measure vital cooling water levels that the NRC warned could give false readings during shutdown (Wald, 1993).

Because incidents and near-accidents recur with disturbing regularity, the public is subjected to a steady stream of discomforting reminders of the dangers of nuclear power. Yet such recurring mishaps should not be unexpected. Charles Perrow (1984) has argued persuasively that nuclear power falls in a class of hazardous technologies that are so complex that their management is inherently difficult and safety-related incidents are unavoidable. This becomes of even greater concern when the potential exists for a catastrophic accident. Perrow points out that nuclear power technology is characterized by many interdependent links that can affect one another but that do not necessarily operate in direct sequences and that may respond in unexpected ways. This may at times make plant functions incomprehensible to their operators, as in the TMI accident where operator error played a large role. Small errors can have large consequences. In 1979, a worker changing a light bulb in the control panel of the Rancho Seco plant accidentally dropped it and short-circuited other controls, causing the reactor to shut down unexpectedly (Perrow, 1984: 44). In the process the pressure vessel was allowed to cool too quickly, thereby increasing the risk that it could crack. Perrow notes similarly that the construction errors and equipment failures that are unavoidable in large engineering enterprises may have drastic consequences in technologies that have catastrophic potential. His analysis points to why mishaps occur routinely, why it is difficult to assure the safety of such systems, and why even seemingly small incidents have the potential of leading to a serious accident.

From the early days of the industry, the AEC required reactors to be designed and built to withstand the effects of a "design basis accident." This was defined as the most severe accident the agency believed to be credibly possible. Because the agency took the position that an accident that could cause severe core damage — let alone a significant offsite release — was "incredible" (Walker, 1992: 384-395), the regulatory process did not require that reactors be designed to withstand such an accident. Similarly, no serious effort was devoted to studying possible mechanisms that could cause such a disaster, despite concerns that began to arise in the mid-1960s (Okrent, 1981, Chs. 8, 11).

It was not until after TMI — by which time the design and principal construction of the present generation of U.S. plants had already largely been completed — that the NRC conceded that accident sequences that could cause core melts and severe offsite consequences were in fact credible. However, the agency still made no changes to its fundamental design requirements for reactors and their containments, which even today remain based on designing against accidents less severe than core-melts. (See 10 CFR 50, Appendix A for the pertinent regulations.) Rather, the agency initiated several programs with regard to severe accident issues, including regulatory efforts to identify and correct vulnerabilities in existing reactors and research programs to attempt to gain a better understanding of the factors and phenomena that could lead to such accidents, e.g., the TMI Action Plan (1980), a Policy Statement on Severe Reactor Accidents (1985); an ongoing effort to identify and resolve a list of Generic Safety Issues; and a research program investigating many phenomena known or suspected to contribute to severe accident risk. But many of these NRC efforts have been controversial. For example, a 1985 National Research Council report criticized NRC's poor management of its nuclear safety research program (National Research Council, 1985); Commissioner Asselstine wrote a stinging dissent to the severe accident policy statement (50 Fed. Reg. 32145ff, August 8, 1985); and a 1989 Congressional report detailed NRC's failure to require many of the post-TMI modifications recommended almost a decade earlier (Wald, 1989). Numerous important technical issues pertaining to severe accidents remain unresolved, particularly with regard to phenomena and sequences that could affect accident progress. (For a representative list, see NRC, 1994: 184-206). The net result is that the NRC's research and regulatory programs regarding severe accidents have not produced the unqualified safety assurances that the public seeks. The possibility of disaster still looms large in the public mind.

With the threat of a catastrophic accident inherent in the nature of current nuclear power plants, a key question arises: how and why did the United States make such a large, rapid, and costly commitment to commercial nuclear power without requiring safer designs in the selection and development of the technology? The answer lies in the unique history of the origins of the nuclear power industry.

Nuclear Power Development and Regulation: The Early Years

The unusually rapid development of nuclear power was the result of broader social and political concerns that dominated the postwar era. These forces combined to create an unprecedented "technology push" to develop a nuclear power industry, with the federal government doing most of the pushing. The result was that safety considerations were overshadowed by the more compelling concerns of the time.

The industry was a direct offshoot of the nation's growing nuclear weapons program, which through the 1940s was what "nuclear" largely meant. The Atomic Energy Act of 1946 created the civilian Atomic Energy Commission (AEC), the nation's first nuclear regulatory agency, and the Joint Committee on Atomic Energy (JCAE), which was charged with all oversight and legislation.[2] Both were responsible for military as well as civilian programs, and in keeping with the Cold War era, both were granted unusually broad powers (Rolph, 1979: 22). Most important, language regarding public health and safety was entirely absent from the 1946 Act (Rolph, 1979).

At the outset, nuclear power "clearly was a secondary goal" (Walker, 1992: 3). It wasn't until the early 1950s that the federal government began actively to promote direct private sector participation in nuclear energy development.[3] The passage of the 1954 amendments to the Atomic Energy Act opened the way for a domestic nuclear power industry by mandating the AEC's promotion of private sector development and by substantially modifying the government's monopoly over the technology (Rolph, 1979). But the legislative history is devoid of any discussion of the nature and magnitude of the potential threats to health and safety (Green, 1981). Even in the congressional hearings and debates on

---

[2] The JCAE for many years routinely promoted nuclear power and shied away from tough oversight (Green and Rosenthal, 1963).

[3] The precipitating event was Eisenhower's December 1953 Atoms for Peace proposal, which itself was largely driven by broader geopolitical and economic concerns. See Walker, 1992: 8-9; and Clarfield and Wiecek, 1984.

the legislation there was little discussion of safety (Mazuzan and Trask, 1979: 37-38).

The changes in the 1954 Act were predicated on the notion that government and the private sector would join forces to develop nuclear power as a domestic energy source. But the technology's potential military applications, the Cold War environment, and the burgeoning growth of what was to become a vast nuclear weapons program resulted in the extremely heavy and active involvement of the federal government, and from early on also created unusually close ties between the industry and its regulatory agency. Unexpectedly high costs in the weapons program created additional pressure on federal officials to develop a civilian power industry that could help justify the government's considerable expenditures (Del Sesto, 1979: 48-49) and in turn also made it apparent that substantial government subsidies would be necessary to assist the early phase of commercial development.

One indication of the lengths to which government was willing to go to induce industry to join in the enterprise was the 1957 Price-Anderson Act, which shields nuclear utilities, vendors, and suppliers against liability claims in the event of a catastrophic accident by imposing an upper limit on private sector liability. Without such protection, private companies were unwilling to become involved (Clarfield and Wiecek, 1984: 197-200; Mazuzan and Walker, 1985: 104-107, and references therein). No other technology in the history of American industry has enjoyed such a continuing blanket liability protection. The Act also sends a disturbing mixed message about nuclear safety: the same probabilities of a catastrophic accident that the industry argues are low enough to merit putting the public at risk are higher than the industry itself is willing to bear in terms of accepting unlimited liability exposure.

The AEC was charged originally with both the promotion and regulation of nuclear power. This created an inherent conflict of interest from early on, which was exacerbated by the close relationship and largely shared agenda between the agency and private industry. This would eventually lead to a loss of public trust, and in 1974 Congress responded by giving responsibility for civilian nuclear power regulation to the newly formed Nuclear Regulatory Commission (NRC). Unfortunately, despite Congress' intent, the new agency continued to favor promotion over regulation (as will be described in greater detail below) (GAO, 1980: 6).

In summary, the heavy federal involvement and broad authority conferred by the Atomic Energy Act and its amendments may have been appropriate for a Cold War program that remained predominantly military in its mission. But when the program emphasis shifted to include the promotion and development of a commercial technology they were unprecedented. The subsequent effort by the federal government to encourage private corporations to join it in a long-term program to promote, research, develop, and commercialize a new and untried energy technology — one which by its very nature involved very large quantities of highly dangerous and toxic materials — remains one of the most unique and unparalleled aspects of the nuclear enterprise. The large financial subsidies and protective liability legislation were indicative of the strength of the government's will to proceed. Taken together, these factors served to force technology development while simultaneously insulating it from public policy debate, the constructive crucible of the marketplace, legal constraints, and the illuminating struggles of pluralistic politics that can serve to define the choices and values at issue.

Technology Development and Reactor Safety

In the early technology development phase of the U.S. nuclear power industry, when other dominant concerns created a headlong momentum to proceed, safety was a distinct afterthought. For a technology with the potential for a large-scale disaster, this seems in retrospect a rather glaring omission. No purposeful and conscientious effort was made to select or develop reactor technology that offered clear safety advantages nor were questions related to safety fully explored. In fact, "there are no records of sustained deliberation over safety and [public] acceptability in the available documents of the archives of the AEC; nor does the detailed account by the AEC's chief historian indicate any search in the 1950s for reactors that would be particularly safe or well suited to widespread use in society" (Morone and Woodhouse, 1989: 40-41). One important consequence of this early failure to consider safety and public acceptance was to constrain unnecessarily the range of possible technology choices.

Although safety was not a prime consideration in evaluating reactor design alternatives, the AEC did take several steps that were driven by safety-related concerns. Early reactors were required to incorporate conservative design features such as redundant systems and emergency cooling capabilities (AEC, 1950). The combination of these features along with remote siting and other elements such as licensing and construction reviews, training, and requirements for secondary safety systems became known as the "defense-in-depth" approach to nuclear reactor safety (Nuclear Energy Policy Study Group, 1977: 232-233), which even today remains the fundamental safety philosophy of the industry and its regulators.

In 1947 an independent advisory panel, the Reactor Safeguards Committee, was formed to advise the Commission on safety issues. (In 1953 this committee merged with a siting panel and the new body was named the Advisory Committee on Reactor Safeguards, the "ACRS.") On the advice of the Committee, early reactors were sited away from heavily populated areas, and from the late 1950s on, were designed to include containment structures to provide a barrier against the release of radioactive materials to the environment in the event of an accident (AEC, 1950; Rolph, 1979: 60-61; Morone and Woodhouse, 1989: 70). But while containments remained a standard requirement of the basic safety strategy, the remote siting requirement was not retained for long.

There were several problems with remote siting. Sites that had been remote when reactors were licensed became more populated over time; regions in which nuclear power was economically attractive had a limited number of remote sites; and siting a reactor remotely resulted in substantially increased utility costs for transmission of the power to the markets where it was needed. As early as 1956 the AEC began to edge away from the policy of remote siting (Openshaw, 1986: 201ff). By the mid-1960s, the pressure to drop remote siting had become intense. Utilities and the nuclear industry lobbied hard for permission to site reactors close to populated areas — in some cases, actually within metropolitan areas (Okrent, 1981: 145-146).

Another factor affecting safety also came into play as the industry grew: utilities began ordering large numbers of much bigger reactors. Through 1964, a total of six U.S. commercial reactors were operating and only thirteen reactors had been ordered (DOE/EIA, 1993: 241; DOE/EIA, 1982: 10). However, beginning the following year the utility industry's enthusiasm for nuclear power generated a spate of new orders: between 1965-1970, utilities ordered ninety-four reactors; and in 1971-1974, the industry placed orders for an additional one hundred thirty-one plants (Openshaw, 1986: 193). As the orders flowed in, the size of the ordered plants grew dramatically. Because costs were higher than had been anticipated, the industry sought economies of scale as a way to lower costs and thus compete effectively with fossil fuel generators. By 1968, utilities were ordering reactors that were two to six times larger than any reactor then in operation (Bupp and Derian, 1981: 72-74). As NRC Commissioner Peter Bradford described it, "an entire generation of large plants was designed and built with no relevant operating experience" (UCS, 1985: 5).

For the safety debate, the rapid scale-up in reactor size was a pivotal event. Larger reactors necessarily use more fuel and create more heat, and bigger cores contain much larger quantities of dangerous fission products. After the TMI accident, a ranking staff member of the NRC explained the fundamental safety difference between larger and smaller reactors to the President's Commission on the Accident at Three Mile Island: "One obvious difference for very small plants is there is no chance of the core melting down and going through the bottom of the reactor vessel because it doesn't have that much energy in it" (President's Commission on the Accident at Three Mile Island, 1979b: 10, note 50). Moreover, in larger reactors the decay heat at shutdown is substantial and requires more cooling for longer periods than in smaller plants. In fact, the decay heat at shutdown for the large reactors being ordered in the mid-1960s was equal to the full-power heat levels of Shippingport, the first fully commercial U.S. nuclear plant (Morone and Woodhouse, 1989: 79-80, note 20 citing JCAE hearings).

Members of the ACRS became uncomfortable with the safety-related implications of the significantly larger cores in the new generation of reactors, which called into question previous calculations and assumptions about what could happen in the event of a serious accident. The larger cores raised new issues with regard to the possible course of severe accidents, including core melts that could breach the containment. Thus containments might no longer be the impregnable barrier to a large release that they had been generally thought to be (Okrent, 1981, chapters 8 and 11). For the same reasons, the AEC's ongoing relaxation of the requirement for remote siting began to cause the ACRS greater concern.

All this in turn posed a problem for the AEC. The relationship between the ACRS and the AEC was a complicated one. In 1957, Congress had established the ACRS as an independent body and required the ACRS to review the safety aspects of all license applications (Mazuzan and Trask, 1979: 47). Contentious issues were often discussed in private between the AEC and ACRS, and the ACRS' published documents and public statements were purposely limited in scope to avoid public controversy (Ford, 1984: 87-88). Nonetheless, on some safety-related issues there was tension between the more cautious ACRS and the more promotionally-oriented AEC. These conflicts were frequently resolved by the AEC ignoring what it regarded as "overconservative" recommendations by the ACRS (Okrent, 1981: 27). Thus in 1965-66 when the ACRS began to raise questions about the severe accident implications of large reactors, the AEC was not receptive. According to former ACRS member David Okrent, "The nuclear industry, the development side of the AEC, and the AEC commissioners themselves were opposed to any consideration within the regulatory process of accidents involving core melt or even severely degraded core cooling . . ." (1981: 164). When the ACRS persisted in 1966 by backing its concerns with a recommendation for an extensive severe accident research program, the AEC responded by establishing a task force that decided instead to focus on improved emergency core cooling systems (ECCS) as a way to mitigate the problem. Half of the task force members came from industry (Okrent, 1981: 165ff).

This decision was a landmark event that signified a fundamental shift in regulatory philosophy not well understood at the time. In brief, the AEC turned away from the arguments that (a) the combination of remote siting, containments, and small reactors was sufficient to mitigate any serious offsite consequences, and therefore (b) the probability of a severe accident with major consequences was so low that such an accident "couldn't credibly occur." Instead, the agency adopted a strategy that embraced large reactors, compromised on remote siting, and emphasized *preventing* accidents through increased reliance on engineered safety systems and greater "defense in depth" (Mazuzan and Trask, 1979: 61; Rolph, 1977: 39-40; Morone and Woodhouse, 1989: 79-81). In practice, this meant that the

Commission had embarked on a never-ending search to prove the impossible — namely, that all significant chains of events and causes that could lead to a severe accident with offsite consequences could be foreseen and that "fail-safe" engineered systems could be designed and maintained to prevent them. Former ACRS member David Okrent has written that this change in regulatory approach in 1966 was of fundamental importance. According to Okrent, "[T]his was the beginning of a continuing series of efforts that looked in ever-expanding directions for possible causes of initiating events that could lead to core melt, and sought measures to reduce the probability of such events" (1981: 135).

In hindsight, it is apparent that this strategy was ill-founded. The systems were too complex, there were too many unknowns, and there was too little operating experience, particularly with large reactors. The AEC had opened a Pandora's box of technical controversies that could never be closed. Any perceived significant lapse in defense-in-depth could justifiably and reasonably be attacked as a safety concern: in years to come, the universe of safety issues would expand to include construction practices, equipment design and maintenance, personnel training and procedures, quality assurance programs, and emergency preparedness. Given the system the AEC had adopted, each new problem that came to light had to be assessed separately to determine the seriousness of its implications for safety, and many became matters of concern for the community of technically sophisticated critics that began to emerge in the early 1970s. Critics were also bolstered in this period by the rapid rise of the nascent environmental movement and citizen activism, the passage of the National Environmental Policy Act (NEPA) and the Freedom of Information Act (FOIA), and the federal judiciary's decision in the Calvert Cliffs case requiring a recalcitrant AEC to meet the requirements of NEPA with detailed environmental impact statements (Mazuzan and Trask, 1979: 61-69).

These events set the stage for the first widely publicized technical controversy about the safety of nuclear power: the 1972 hearings on the adequacy of emergency core cooling systems (ECCS). The ECCS controversy marked the critical turning point in the safety and accident debate. The AEC had responded to the ACRS safety concerns about larger reactors by initiating a program in the mid-1960s to investigate the effectiveness of ECCS designs with the objective of requiring improved ECCS as a way to prevent accidents. By choosing to focus on the ECCS instead of core-melt issues directly, the AEC had made the fateful decision to emphasize defense-in-depth and accident prevention as the primary defense against severe consequences from severe accidents rather than retaining smaller reactors and remote siting.

The ECCS controversy also cast the bright light of public scrutiny on the AEC's emphasis on promotion over regulation. The AEC contracted with the national laboratories to do the work on the ECCS research program. Deadlines were missed for important milestones and tests, and the data that did emerge were not reassuring. The AEC issued new interim criteria for ECCS improvements in 1971, but the Union of Concerned Scientists (UCS) responded by publishing technical critiques that were widely circulated. In response, the AEC decided to hold an informal rulemaking hearing in early 1972 in which intervenor groups participated, led by UCS. Industry argued that the AEC's standards were, if anything, too conservative. But as the hearings dragged on far longer than planned, the AEC's position deteriorated as damaging information came to light. The controversy was replete with charges of the AEC's withholding important documents, AEC intimidation of the staff of its contractors at the national laboratories, and profound disagreements among technical experts, including some of the AEC's own contractor staff (Gillette, 1972a-1972d; Ford, 1984). The adequacy of the ECCS remained a point of debate for years to come and was raised in numerous plant licensing hearings. (For a fuller account of these events, see Rolph, 1979; Ford, 1984; Walker, 1992.)

For the first time the public became aware of profound technical disagreements among experts about vital safety matters. Moreover, the AEC was roundly criticized by the national press, which posed the basic question of whether the AEC's first priority was promotion of the industry or protection of the public health and safety. The damage to the AEC's public credibility was fatal, leading to the dissolution of the agency within a few years. On the other hand, the credibility of intervenors was enhanced significantly, and they were widely credited with having exposed AEC efforts to brush aside an important safety question and with forcing the AEC and industry to adopt more effective ECCS safety criteria (see, for example, Cottrell, 1974: 53).

With the rise of environmentalism in the 1970s, the antinuclear movement grew substantially. In 1975-1976, ballot initiatives to control or halt the growth of nuclear power were introduced in eight western states. Although they enjoyed little success at the polls, the controls they sought to impose were sometimes adopted in part by the state legislature, most notably in California. Interventions in plant licensing proceedings increased, often focusing on technical issues related to safety. This widespread popular ferment kept the issue before the public and contributed to growing public skepticism about nuclear power. The accident at TMI marked another milestone, causing public uneasiness about the safety of nuclear power to rise to unprecedented levels which have largely continued to the present day.

The popular movement against nuclear power continued into the 1980s, but began to lose momentum when interest in energy as a national policy issue declined, fewer plants remained in the construction and licensing pipeline, and many of the movement's supporters switched their focus to nuclear weapons and the arms race. Regardless, in the aftermath of TMI there was broad media coverage about nuclear power safety issues, and several contentious plant licensing proceedings (most noticeably Shoreham, Seabrook, and Diablo Canyon) kept accident and safety issues before the public. "Direct action" protests in the form of civil disobedience and sit-ins occurred at some of the contested plants. In 1986, the Chernobyl accident sharply refocused national and world attention on nuclear safety.

Thus ignored at the outset, safety concerns grew as efforts to develop and commercialize nuclear power progressed. To try to address the issue, the AEC commissioned a series of studies that attempted to estimate the consequences and risks of a severe accident.

Severe Accident Risk: Consequences and Probabilities

The first detailed and systematic effort to quantify the possible consequences of a severe nuclear power plant accident was commissioned by the AEC at the request of the JCAE and released in 1957. The study, referred to by its AEC document reference number as WASH-740, was produced by Brookhaven National Laboratory. The study presumed an accident at a 100-200 MW(e) power reactor located near a major city in which roughly fifty percent of the reactor's core would be carried by winds to reach the city environs (Mazuzan and Walker, 1985: 206-207). According to WASH-740, the consequences of such an accident could total 3,400 deaths, 43,000 injuries, and $7 billion in property damages (Walker, 1992: 114-115). Although the probability of the accident was presumed to be "exceedingly low," the lack of reactor operating experience precluded a firm statistical estimate. Needless to say, the publication of these estimates caused consternation in many quarters.

In 1964, the JCAE asked the AEC to update WASH-740 prior to proposing an extension of the Price-Anderson Act. The

AEC turned again to Brookhaven. The assignment was to estimate the consequences of the worst accident that couldn't be demonstrably shown to be false. Lacking a methodology to deal with large uncertainties, the study team decided not to try to estimate accident probabilities. But in response to the larger reactors then being proposed, the Brookhaven team increased the presumed fission product inventory of the hypothetical reactor core and assumed that a core-melt accident would result in the escape of a large quantity of radioactive materials to the environment. In 1965, a draft had been completed which estimated that such an accident could cause 45,000 deaths, 70,000 injuries, significantly contaminate 10,000 to 100,000 square kilometers, and cause $17 billion or more in damages (Walker, 1992: 125; Ford, 1984: 69; UCS, 1977: 3-5).

An industry group to which the draft had been sent for review appealed to the AEC not to release the results, fearing a public relations disaster. Not unexpectedly, the AEC, an influential member of the JCAE who had been shown a draft, and most members of the ACRS were also disturbed by the study results. After much wrangling among the parties, the AEC decided to suppress the study and, in essence, to deny its existence by treating it as a draft that had not been completed (Walker, 1992: 121-131). It was not until 1973 that the study papers were finally made public, under pressure of an FOIA request (UCS, 1977: 148).

In 1972, the AEC decided to commission yet another accident consequence study, but this time the work would be conducted by the Commission itself and contractors it selected. The new study would utilize an emerging methodology known as probabilistic risk assessment (PRA) to estimate not only the possible consequences of a range of reactor accidents, but also the probability of their occurrence. The report was released in 1975, and was eventually referred to by two names: WASH-1400, or the "Reactor Safety Study" (NRC, 1975a). The WASH-1400 study concluded that a credible worst-case accident could cause 3,300 early fatalities, 45,000 early illnesses, 45,000 latent cancer fatalities, 240,000 cases of thyroid disorders over the long term, and $14 billion in property damage (NRC, 1975a: 83, table 5-5). But it also asserted that the probability of such a severe accident occurring was extremely low (on the order of one in a billion), and that accidents with higher probabilities had much less severe consequences. The report's executive summary asserted similarly that the annual risk of death to an individual was a minuscule 1 in 5 billion (NRC, 1975b: 3). Nuclear industry advocates finally had the accident study they had always sought: one that tempered high consequence estimates with calculated estimates of low probability.

WASH-1400 generated a storm of criticism in the years following its release. Several detailed critiques raised numerous questions about the study's assumptions, methodology, calculations, peer review procedures, and objectivity (UCS, 1977; NRC, 1978; Ford, 1984: 164ff). The controversy over the report focused in large part on the executive summary, which critics claimed painted a misleading picture by making unsupportable claims and glossing over or ignoring entirely debatable assertions. This led to a request from Congress to the NRC to empanel a special Risk Assessment Review Group to review the executive summary's portrayal of the study's results (Okrent, 1981: 318). Chaired by Harold Lewis, the group (also referred to as the Lewis Committee) concluded that the executive summary was "a poor description of the contents of the report, should not be portrayed as such, and has lent itself to misuse in the discussion of reactor risks" (NRC, 1978: vii). The group's report also cited problems in WASH-1400's completeness, treatment of common cause failures, and radiation dose calculations. Moreover, the reviewers stated that "we are unable to define whether the overall probability of a core melt given in WASH-1400 is high or low, but we are certain that the error bands are understated. We cannot say how much" (NRC, 1978: vi). On a more positive note, the Lewis Committee noted that PRA methodology was sound for some purposes and deserved wider use in the regulatory process and to identify some safety-related issues (NRC, 1978: vii-ix).

As a consequence of the strength of this and other critiques, the NRC in 1979 was forced to issue a policy statement in which it withdrew any endorsement of WASH-1400's executive summary and accepted numerous criticisms of the full study raised by the Lewis Committee. The NRC explicitly stated that it did not regard as reliable the Reactor Safety Study's numerical estimate of the overall risk of reactor accident (NRC, 1979).

In late 1982 the most recently available study addressing industry-wide accident consequences was released. The study was performed by Sandia National Laboratory for the NRC, and became known as the Sandia Siting Study (NRC, 1982). The study presumed several different accident scenarios, one of which was a core-melt accident with a large release of radioactive materials. The probability of such an occurrence in any given year was estimated as 1 in 100,000, with a very large uncertainty estimate of up to 100 times higher and 1000 times lower (GAO, 1987: 18, n1). The study examined the potential consequences of an accident at a hypothetical 1,120 MW(e) large reactor for 91 separate U.S. reactor sites. The investigators used sophisticated computer models that ran many different scenarios for each site using different combinations of site-specific data for wind directions, precipitation, population density, and so forth. When the study was released, others scaled the results proportionately for reactors smaller than the 1,120 MW reference case. For some sites, the results under worst-case local weather conditions were staggering: for the Salem site in New Jersey, 103,000 early fatalities were estimated; at the Indian Point site, property damages exceeded $300 billion. Averaged (as opposed to worst-case) results for each site yielded lower numbers, albeit still disturbingly large (OTA, 1984: 219).

The GAO released two reports in 1986 and 1987 that used the Sandia data to derive average financial consequences over a range of worst-case accidents at reactor sites for each of the 119 plants then operating. Financial consequences included dollar estimates for injuries, deaths, and property damage (in 1986 dollars). GAO determined that the financial consequences of an accident could range from $15 million to $67 billion, depending on the plant. The consequences for 113 plants were $6 billion or less. The GAO emphasized that both accident probability and consequence estimates were subject to considerable uncertainty (GAO, 1986, 1987). Public interest groups criticized the GAO reports, charging that they had relied on methodologies and assumptions of WASH-1400 that were inappropriate or mistaken, that GAO had accepted nuclear industry and NRC opinion on important aspects without seeking independent review or comment, that "worst case" possibilities had been ignored, and that cost factors were seriously underestimated (Gordon and Knapp, 1989).

In the mid-1980s, the debate shifted to a discussion of "source terms," defined by the NRC as the fraction of the total radionuclide inventory of a reactor at the start of an accident that is released to the environment, including calculations to factor in the energy, height, and timing of the release (NRC, 1990a: 2-3). For several years the industry and the NRC argued with critics about various accident phenomena that could affect source term estimates and thus also estimates of accident risks and consequences. A 1986 NRC report concluded that source terms were highly dependent on the specific construction details of individual plants, and were especially dependent on containment behavior (NRC, 1986a). However, many key questions resisted definitive answers and the focus on source terms faded as the decade ended. (For critics' appraisals of the source term debate, see Sholly and Thompson, 1986. For the NRC's views of the current status of the source term issue, see NRC, 1995.)

The Sandia studies and the 1986 and 1987 GAO reports based on their data were the last major generic U.S. accident consequence studies to date. These studies have posed a thorny problem for nuclear power proponents. Through the early 1970s studies showed that disastrous consequences could occur from a severe accident, but despite assertions that the risk of such an accident was low, no specific probabilities were estimated. WASH-1400 calculated such probabilities, but its methodology raised questions about the accuracy of those estimates. The NRC's next step was to try to address the issue of accident probabilities.

From Estimating Consequences to Estimating Probability

In the early 1980s, the NRC embarked on several new approaches to the issue of accidents and plant safety, largely in response to the TMI accident. Rather than seeking to estimate generalized consequences, these focused instead on setting risk-related objectives and more closely analyzing the probability of accidents at individual plants as new ways to approach the basic questions about safety.

One such approach was the setting of "safety goals," which were adopted in final form in 1986 at the end of a six-year development process. Briefly, there are two NRC qualitative safety goals, accompanied by two quantitative objectives intended to provide numerical guidance or "aiming points" for the qualitative goals. The qualitative goals are (paraphrased) that "individual members of the public . . . should bear no additional risk to life and health" from nuclear power plants; and societal risks should be comparable to or less than those of competing energy technologies. The quantitative objectives are that the risk of immediate death from a reactor accident to an average individual in the vicinity of a nuclear power plant should not exceed 0.1 percent of the sum of the risks of immediate death from other causes; and the accident risk of death from cancer to the population near a plant should not exceed 0.1 percent of the sum of the risk of cancer death from all other sources (51 Fed. Reg. 30028-30033, April 21, 1986).

Exactly how these safety goals and objectives were to be used in the regulatory process was not clear. The NRC statement indicated that cost-benefit analyses were to be used to determine in part whether modifications were warranted at existing plants if their assessed risk exceeded the objectives. It also noted that the Commission would issue additional implementation guidelines to the staff "to use as a basis for determining whether a level of safety ascribed to a plant is consistent with the safety goals." However, as of the end of 1995 specific guidelines had not been issued. The Commission proposed initially as one guideline that the "overall mean frequency of a large release of radioactive materials to the environment from a reactor accident should be less than 1 in 1,000,000 per year of reactor operation" (51 Fed. Reg. 30031). In other statements the NRC has stated that the quantitative objective for a core melt or severe core damage accident (regardless of whether it results in a release to the environment) should be less than 1 in 10,000 per reactor per year (NRC, 1990b: 3). The NRC indicated in its safety goals policy statement that probabilistic risk assessment was one tool — albeit not a final or definitive one — to be used to judge whether or not a plant was meeting the safety goal quantitative objectives.

The NRC safety goals have been criticized on the grounds that they were less ambitious or comprehensive than they might have been, and that some aspects initially followed more closely an approach suggested by industry rather than that recommended by the ACRS (Okrent, 1987; see also separate views of Commissioner Asselstine, 51 Fed. Reg. 30032-30033, August 21, 1986). Others complained that they were not specific enough (see, for example, Commissioner Bernthal's comment at 51 Fed. Reg. 30033).

The probability of a serious accident was also addressed by the NRC in 1985 testimony to Congress. The NRC asserted that the probability of a severe core damage accident for one hundred operating reactors over the next twenty years was 45 percent. Commissioner James Asselstine dissented, saying that because the uncertainties of the estimates were so great the actual risk was somewhere between 6 and 99 percent and could not be pinned down with any greater accuracy (U.S. Congress, 1985: 378). The NRC staff later reduced their estimate to 12 percent, based on an extrapolation of the early results of five plant-specific studies that suggested the risk of a severe core-damage accident was lower than earlier believed (NRC, 1986b).

The accident at TMI reinforced the public's concerns that despite oft-repeated industry assertions to the contrary, things can go very wrong at nuclear power plants. Public opinion plummeted despite industry's somewhat tortured claims that the TMI accident proved that reactors were safe because emergency systems had worked by preventing a worse accident. A technical analysis of the TMI accident completed in 1993 revealed that it had indeed been a core-melt accident in which a substantial portion of the fuel had melted.

Following Chernobyl there were numerous industry assurances that an accident with consequences as severe could not occur in the United States owing to the design differences between Soviet-constructed and U.S. plants. But once again, there was disagreement among the experts. Commissioner Asselstine testified before Congress in 1986, stating that "[W]hile we hope that their occurrence is unlikely, there are accident sequences for U.S. plants that can lead to rupture or bypassing of the containment in U.S. reactors which would result in the off-site release of fission products comparable to or worse than the releases estimated by the NRC staff to have taken place during the Chernobyl accident" (U.S. Congress, 1986a: 38).

In a further effort to characterize severe accident risks, in 1990 the NRC published the final version of NUREG-1150, a study begun several years earlier by Sandia National Laboratories which was comprised of five plant-specific PRAs for individual, representative plants. A principal objective of NUREG-1150 was to derive plant-specific quantitative risk estimates (including those for core-melt and large release probabilities) in order to update the estimates from WASH-1400, identify accident sequences that presented the greatest risk, and identify changes in procedures or equipment that might help reduce that risk (NRC, 1990b).

As early results began to come in from the NUREG-1150 program, the NRC decided that plant-specific PRAs could better define risk estimates and provide information about causal events and weaknesses in individual plant systems which could be used to identify safety improvements. As a consequence, in 1988 the NRC required all individual reactors to conduct plant-specific PRAs (NRC, 1988). By the end of 1995, all plants were expected to have completed these "individual plant examinations" (IPEs). The NRC permitted the IPEs to use non-PRA estimation techniques for some important classes of external initiating events such as fires and earthquakes, and also relied on expert judgments in some instances where important data were not available. This added to the already large uncertainties to which PRAs are subject as an artifact of the methodology itself. Moreover, many observers — including at least one NRC Commissioner — have expressed concern that the numerical results of PRAs will be regarded as "bottom line" point estimates of risk (one of the principal criticisms of WASH-1400), despite their wide uncertainty range (GAO 1985: 68). A cursory look at the literature suggests that some in industry are indeed pointing to low plant-specific PRA core-melt and large release frequency estimates as evidence that plants are "safe enough" and that quantitative safety targets are largely being met.

Steven Sholly of MHB Technical Associates has compiled the PRA risk estimates for all IPEs completed as of 1994. His unpublished analysis shows that for core melt, the average estimates (i.e., the mean point estimates, regardless of the error band) for most plants are hovering in a range just below 1 in 10,000 (the average is actually 6-7 in 100,000).[4] For 109 reactors, this amounts to a 15 percent chance of a core-melt over the next 20 years – a number quite close to the Commission's 1986 estimate of 12 percent (Sholly, 1995). While a core melt would not necessarily cause significant offsite releases, it would surely constitute a major financial event for any utility whose reactor should suffer the accident. The experience at TMI indicates it would also be traumatic for affected states and communities.

More important, Sholly's analysis reveals that roughly three-quarters of the plants are not meeting the large release target of 1 in 1,000,000 per reactor year – some by as much as a factor of 10 or more. Multiplying this probability by 109 plants for 20 years results in a 1.7 percent chance of a large release to the environment over the next 20 years. This is hardly an insubstantial number for an accident with significant offsite health and economic consequences, particularly given the significant uncertainties inherent in the analytical techniques and with regard to associated factors such as core-melt phenomena, human errors, and common-cause failures.

It should be apparent from the above that the NRC's approach to what constitutes the answer to the question "How safe are nuclear plants?" has changed frequently over the past twenty years. PRAs and their application to individual plants are the approach currently receiving the greatest emphasis. But it should be equally apparent that PRAs are not providing the definitive reassurance the public seeks about nuclear power's safety and

---

[4] Sholly recognizes that it is inappropriate to use PRA point estimates as "bottom line" indicators of safety. However, because many in industry continue to use PRA point estimates in that manner, he sought in part to analyze the data as given to put such claims in perspective, i.e., the industry is not meeting its own proffered objectives.

risks, and that public opinion about nuclear power is not changing significantly as a result.   Rather, it appears that the accident consequence studies conducted since 1957 have been far more influential in shaping public sentiment about nuclear power. The public remains aware that a catastrophic accident could occur, but has not been reassured that the probability of disaster is sufficiently low that in the public's mind nuclear power plants are to be regarded as "safe."

There is common agreement that the probability of a severe accident is not so high as to make such events common; however, disagreements remain about just how low the probabilities actually are, about the validity of the methodologies used to derive the estimates, and about how low is "low enough."   Even if one accepts the claim that the probability of accident is indeed low–disregarding for the moment the disagreements about just how low–why does the public persist in regarding nuclear power as unsafe? To answer that question requires an understanding of the factors that underlie people's opinions about risk.

Risk Perception and Assessment: The Public Versus Professions

Risk assessment seeks to define and rank risks by amassing statistical data and crafting cost-benefit analyses. Risk perception is the field of study of social scientists. It explores how people perceive risks — what beliefs are held, the bases for those beliefs, and how and why the public comparatively rates different kinds of risks. On the topic of nuclear power, a wide gulf exists between the vantage points of the public and risk assessment experts. Despite technical studies that assert that the probability of a severe reactor accident is low, numerous surveys have shown that the public remains very deeply mistrustful and uneasy about nuclear power. People's perceptions of the risks of nuclear power (and other hazardous technologies) are largely unresponsive to the claims of technical risk assessments (Slovic, 1993).

Risk assessment experts — especially those allied with the industry — often express bemusement and frustration about the public's mistrust of nuclear power. They cite statistical data that demonstrate that in the normal course of things, the annual risk to the individual of many activities, such as smoking, motor vehicles, handguns, alcohol, and surgery, far outweigh the annual risk from a nuclear power plant accident. Interestingly, there is empirical evidence that risk assessment experts and laypeople alike rate the relative hazards of many such risks similarly (Slovic et al, 1982). But nuclear power is the hazard for which the widest divergence in such ratings often occurs between experts and the public. In the Slovic et al study (1982), risk assessment experts rated the hazards of nuclear power quite far down on a list of comparative risk of different technologies; members of the public rated the risks of nuclear power near the top. While some may attribute this disparity to the public's poor comprehension of nuclear power technology, on closer examination the public's unease seems much more reasonable than at first glance.

Social scientists interested in risk perception have sought to identify the traits that lead people to regard hazards as high in risk. Survey data have revealed that people rate technologies or activities as particularly hazardous in proportion to the degree to which they regard the technologies and/or risks as (a) not voluntary, i.e., risks to which one is exposed not by personal choice; (b) imperceptible to those exposed to the risk or unknown to science (e.g., radiation and some chemical carcinogens); (c) outside the control of the person who may be exposed to the risk; (d) involving a technology that is unfamiliar or new; (e) presenting the possibility of a severe or fatal consequence; (f) posing the threat of catastrophic and widespread consequences; (g) invoking a fear response, or "dread"; and (h) not perceived as offering compensating benefits (Slovic, 1991).

The public consistently rates nuclear power near the extreme negative ends of the above spectrums, and studies reaching similar conclusions have been replicated in many different countries (Slovic, 1991, and references therein). There are other contributing factors that may also run deep in the public consciousness. Kirk Smith, among other observers, has suggested that the public's consistently negative ratings of the risks of nuclear power are reflective of the industry's unique connection with the history, technology, and imagery of nuclear weapons (1988: 62):

> [One] reason that helps to explain why nuclear power is seen differently from other technologies lies in its parentage and birth. Nuclear energy was conceived in secrecy, born in war, and first revealed to the world in horror. No matter how much proponents try to separate the peaceful from the weapons atom, the connection is firmly embedded in the mind of the public.

Regardless of precisely which factors account for most of the public's mistrust, Paul Slovic, one of the leading researchers on the subject, has written that "nuclear power has a special distinction in the perception literature — it is, to date, the technological hazard with the most negative and most problematic constellation of traits. It stands apart in having qualities that make it fearsome and hard to manage socially and politically" (1991: 2).

The catastrophic potential of nuclear power appears to be a heavily influential factor. This is surely one of the distinctive ways in which nuclear power differs from most modern technologies and from all other energy technologies. From a cultural and psychological perspective, accidents that cause the deaths of hundreds of people in a very short period of time (such as airplane crashes) affect the public consciousness differently than several hundred accidents distributed over time in which one or two people die — a fact often ignored in technical risk assessment. Charles Perrow has commented on this difference in viewpoint (1984: 308-310):

For some economists and risk assessors . . . there is no difference between the death of fifty unrelated people from many communities and the death of fifty from a community of one hundred. Social ties, family continuity, a distinctive culture, and valued human traditions are unquantified and unacknowledged. Fifty thousand highway deaths a year are equivalent to a single catastrophe with fifty thousand casualties for these experts, and they deplore the fact that the public protests nuclear plants and estimates highway deaths to be only half of what they are.

For thirty-five years the public has been the recipient of conflicting messages: (1) the probability of a severe nuclear power plant accident is extremely low, (2) experts disagree about numerous safety-related technical issues, and (3) if a catastrophic accident should occur, many thousands of people could die and thousands more could be injured. Because even industry proponents have been forced to admit that a truly severe accident, however unlikely, could cause catastrophic consequences, the public remains discomforted. Some proponents nonetheless argue that the public is "irrational" and mistaken because it does not believe or behave in accord with what these proponents claim to be "objective" facts and probability estimates, but such arguments belie their failure to understand the broader, powerful cultural and psychological forces related to hazardous technologies that are, in the end, more influential in shaping the public's views. Rather, the public's response to a technology that suffers "normal accidents" as a routine occurrence — and which over longer time frames may be expected to result in some accidents that are catastrophic — seems rational and quite well-grounded indeed.

Slovic, for one, concurs that the public is neither ignorant nor irrational and he cautions against spending "[more] billions of dollars pursuing a path that is destined to lead to failure . . . Public relations efforts won't create trust. Aggressive and competent government regulation, coupled with increased public involvement, oversight, and control, and a 'trouble-free' performance record might" (1991: 8). Unfortunately, impartiality, accessibility, and aggressiveness have not been hallmarks of the regulation of the nuclear power industry. The failure of regulation was one of the prime reasons why the public began in the 1970s to doubt the industry's safety. Yet, instead of reassuring the public about safety by opening up the process, in recent years the NRC has instead acted to exclude the public even more so, and thus risks exacerbating their concerns.

Promotional Regulation and Public Access to the Regulatory Process

The congressional decision in 1974 to divide the AEC into two separate agencies (NRC and what was soon to become the Department of Energy, or DOE) was intended to solve the inherent conflict of interest between promotion and regulation. In essence, the AEC was perceived — with considerable justification — as having lost its commitment to protect the public; instead, the agency was favoring the interests of the very industry that it was charged with regulating. Senator Abraham Ribicoff, who chaired the Senate Committee that proposed the 1974 reorganization, echoed the prevailing sentiment when he explained to his colleagues that "[I]n fact, it is difficult now to determine . . . where the Commission ends and the industry begins" (cited in Mazuzan and Trask, 1979: 86).

The majority of the new NRC's staff and top managers, including the new commissioners, consisted of holdovers from the old AEC. Rather than starting fresh, it meant that the new agency started up with the same people and prejudices that the legislation abolishing its predecessor had been designed to correct. The result was a lost opportunity for change. According to a 1980 GAO report: "The [new] Commission's failure to exercise the opportunity for a searching reappraisal of the direction and approach to nuclear regulation led to continued acceptance and perpetuation of AEC's regulatory principles, priorities, and programs" (GAO, 1980: 46). This is not to say that there were no changes at all. But a strong argument can be made that like its predecessor, the NRC has consistently favored the interests of industry and just as consistently has sought to hamper or deny public access to the regulatory process. Moreover, since the mid-1980s this trend has worsened noticeably. The NRC has sponsored legislation and enacted regulations that place unprecedented new barriers before public participation. The record is replete with instances of the NRC's failure to pursue tough regulation while simultaneously remaining unduly responsive to the industry's concerns.

In plant licensing hearings, the Commission has routinely resisted public participation; indeed, the Commission has gone to considerable lengths to frustrate it. Intervenors cite many examples, including the following: (a) when a utility has sought a construction license, the NRC has set the timing of the hearings such that intervenors must gather information and present safety-related concerns before essential utility documents and NRC staff reviews of them become available (Glitzenstein, 1994a: 159; UCS, 1985: 67); (b) "rules are frequently interpreted narrowly to exclude unfavorable evidence and disregard serious safety issues" (UCS, 1985: 68); and since 1981, the NRC has increasingly acted to cut off public hearings by exercising its authority to enter its licensing boards' proceedings directly and stop inquiries into safety issues (UCS, 1985: 70; see also Asselstine, 1987: 259).

In general, intervenors have encountered hostility and procedural roadblocks raised by the NRC and its licensing boards. Despite these obstacles, intervenors have made positive contributions and have sometimes succeeded in forcing the review of important safety issues by raising their visibility (Rogovin, 1980: 143-144; Rolph, 1979: 123). However, for the most part meaningful public participation in the licensing process has seldom been a reality (e.g., see Del Sesto, 1979: 122-136). This situation was recognized by the NRC's own Special Inquiry Group constituted in response to the TMI accident, whose report bluntly stated, "[I]nsofar as the licensing process is supposed to provide a publicly accessible forum for the resolution of all safety issues relevant to the construction and operation of a nuclear plant, it is a sham" (Rogovin, 1980: 139).

Industry dissatisfaction with even limited access to the process led to the most systematic attempt to curtail public participation via what has been euphemistically called "licensing reform." In the mid-1950s Congress mandated a two-stage licensing process for the construction and operation of nuclear power plants, and provided that the public would have a right to participate in hearings in both stages. These procedures were established to ensure public confidence and to counterbalance the broad powers and privileges granted to the federal government and the industry (Glitzenstein, 1994a: 157). The process has been observed for all plants licensed to date. But in 1989 the NRC issued two new sets of regulations that changed the longstanding licensing arrangement. The first significantly limits the ability of intervenors to raise environmental and safety issues in the licensing process and also limits their ability to pursue issues that are admitted. The second did away with separate construction and operating license hearings by authorizing the Commission to issue a "combined license," and severely restricted the issues that intervenors could raise prior to a plant receiving NRC permission to operate (Glitzenstein, 1994a: 165-72). In 1992, the Commission, along with DOE and the industry, succeeded in convincing Congress to legitimize "one-step" licensing with a provision in the 1992 Energy Policy Act (Glitzenstein, 1994a: 177-182; see the text of the bill, H.R. 776, Section 2801ff).

Advocates of "licensing reform" have pushed through related measures as well. In 1989, the NRC adopted "site banking" regulations that permit a utility to seek approval for a nuclear plant site without either seeking a construction permit or submitting a reactor design. The permits are valid for up to twenty years (plus a possible twenty-year extension) and the utility may seek approval for emergency plans without ever filing an application to construct a plant. Another regulatory change permits applicants to seek via a generic rulemaking a "standard design certification for essentially complete nuclear power plant design" good for a period of fifteen years. This will effectively bar intervenors from raising most safety and environmental issues when a specific plant using a certified design is actually proposed (Glitzenstein, 1994a: 172-74).

The NRC backtracked earlier on nuclear power plant emergency planning. States and local governments have the primary responsibilities for planning for and providing the on-the-ground response to a major nuclear power plant emergency. The President's Commission on TMI recommended specifically that operating licenses should be granted only when a state emergency plan had been approved and effective coordination with local officials had been demonstrated (President's Commission on the Accident at Three Mile Island, 1979a: 76). In the mid-1980s there was strong state opposition to the licensing of the Seabrook and Shoreham nuclear plants on the grounds that no realistic emergency plans were possible for those reactor sites. The NRC responded by changing its regulations to permit it to grant an operating license by approving emergency plans drafted by the utility instead (see 52 Fed. Reg. 42078, November 2, 1987).

As for plants in operation, the public has virtually no opportunity to raise challenges with regard to health, safety, or environmental issues. The public's recourse is primarily through "show-cause petitions" under Section 2.206 of the NRC regulations, which provides that any person "may file a request [to] modify, suspend or revoke a license." The decision to grant a hearing is left to the sole discretion of the NRC, with virtually no opportunity for meaningful judicial review. Between 1981-1991, the NRC granted only two public hearings (both in the early 1980s) in response to a total of 321 show cause petitions (Glitzenstein, 1994b: 194). Actions the NRC does take in response to show cause petitions often involve private discussions with utilities. Moreover, the details of NRC investigations and their outcomes are not made public, and thus there is no way to judge independently whether the agency has sufficiently addressed the problems identified (Glitzenstein, 1994b: 195-196).

The AEC was dissolved because it was perceived as unduly favoring the industry it was charged with regulating. The NRC appears to have fallen into the same trap. A 1987 Congressional report entitled "NRC Coziness With Industry" concluded that the NRC "has not maintained an arms length regulatory posture with the commercial nuclear power industry . . . [and] has, in some critical areas, abdicated its role as a regulator altogether" (U.S. Congress, 1987: 3). To cite just three examples: a 1986 a Congressional report found that the NRC staff had provided valuable technical assistance to the utility seeking an operating license for the controversial Seabrook plant (U.S. Congress, 1986b: 6). In the late 1980s, the NRC "created a policy" of non-enforcement by asserting its discretion not to enforce compliance with license conditions; between September 1989 and 1994, the "NRC has either waived or chosen not to enforce regulations at nuclear power reactors over 340 times" (Riccio, 1994: 1). Finally, critics charge that the NRC has ceded important aspects of its regulatory authority to the industry's own Institute for Nuclear Power Operations (INPO), an organization formed by nuclear utilities in response to the Three Mile Island accident (Utroska, 1987; Riccio and Freedman, 1993b; Simpson, 1994).

Regulation is unavoidably a political process that functions well only when the public has confidence in its fairness (Rolph, 1979). The net result of the NRC's unprecedented attack beginning in the late 1980s on public access to the regulatory process may well turn out to be just the opposite of what is hoped for. Public interest attorney Eric Glitzenstein has observed (1994a: 156):

> There are, at present, enormous legal and practical barriers to effective public involvement in NRC proceedings. If these barriers are not removed or at least lowered — either through legislative action, the NRC's own initiatives, or a combination of both — the public will have little meaningful input into Commission decision making in the future. . . Proceedings which merely offer a pretense of public involvement will accentuate, rather than assuage, public suspicion and skepticism concerning the risks of nuclear power.

Democracy and Nuclear Power: The Social and Political Dimensions of Technology Choice

Regardless of one's personal predisposition for or against nuclear power, there can be little doubt that the industry will continue to fight an uphill battle without public support. As noted at the outset, there is broad agreement that the lack of public acceptance is one of the foremost obstacles blocking a new plant order. Proponents who complain that the public doesn't understand the technology, or somehow has been misled, are missing the point. They only reveal their long-standing difficulty in comprehending that the public looks at risks in a manner quite differently from the way they do.

Nuclear power advocates who insist that their quantitative safety and risk analyses demonstrate that in order to solve our energy problems (as they see them) we must forge ahead on the nuclear path fall into yet another trap, what may be called a "technocratic fallacy." To wit: if scientific and technical analyses seem to support the arguments for the implementation of a technology, this somehow becomes a reason why such a decision *ought* to be taken. But because opponents and the public do not share the technologist's world view does not mean that their preferences and viewpoints should not be accorded equal weight.

What is really happening is a classic case of "talking apples and oranges." As described above, the public's discomfort about the risks of the technology is neither irrational nor without cause, and opponents' criticisms and concerns regarding safety are not without foundation. Certainly there are narrow technical disagreements that in some cases can be resolved by simply coming up with more facts. But the broader dispute arises in part because in important ways the three parties — proponents and their technologist allies, opponents, and the public — view the world through different lenses and end up speaking different languages. The real roots of the conflict are not to be found in narrow technical issues, but rather in differences in cultural attitudes, belief systems, political preferences, and broad social ideologies. Steven Del Sesto (1980) has argued cogently that even conflicting congressional testimony on nuclear power safety issues can be better understood as a contest between different ideological viewpoints rather than as narrow technical disagreements.

The policy opinions of "scientific experts" — regardless of whether they are nuclear power advocates or opponents — are, in the end, subjective judgments about what society *should* do. What are often characterized as mere factual disputes turn out instead to involve complex matters of social and moral values, cultural differences, and alternative visions of the future. This distinction is vital but seldom acknowledged. Once it is recognized, it should be equally apparent that the discussion is really about the process of democracy – the question of how, in a democratic society, such decisions are to be made. Cora Bagley Marrett, who served on the Kemeny Commission, has eloquently described this situation (1984: 324):

The public bids for involvement because it recognizes that the nuclear debate concerns more than just engineering problems. It involves such matters as the level of risk that is and should be acceptable in a society. Many of these problems cannot be solved by careful experimentation and detailed statistical analyses. Rather, they must be recognized as questions of values and ideals, and be dealt with in the political theater instead of the scientific laboratory. When the public seeks involvement in the debate on nuclear power, its aim is not to fashion control rooms or weigh the relative merits of one reactor design over another; it is to have the experts and elected officials recognize that complex technologies raise social as well as technical problems in which citizens clearly have an interest.

This goes to the heart of the public's concerns about nuclear safety and profound discomfort about nuclear power. Despite the ardent wishes of industry proponents that things were different, the future of nuclear power may well depend as much or more on social and political concerns as on technical issues. This is neither a recent development nor a rhetorical assertion. Freudenberg and Rosa wrote in 1984 that this view "has become conventional wisdom in the energy policy establishment" (p. 344), citing as supporting examples five of the most widely influential energy policy studies from 1979-1980. Harold Green, a former AEC attorney who later represented intervenors, put it more plainly (cited in Rolph, 1979: 101):

No elite group of experts, no matter how broadly constituted, has the ability to make an objective and valid determination with respect to what benefits people want and what risks people are willing to assume in order to have the benefits.

For now, the public has spoken loud and clear. If the American people do not change their views about the safety and acceptability of the enterprise, and if they are not excluded from the procedural forums in which they may pursue those views, nuclear power will likely continue to founder in the face of widespread mistrust. But industry proponents who seek to advance their cause via the strategy of forcing the public out of a meaningful role in the regulatory process have enjoyed considerable success in recent years. If that momentum continues, the system will be closed even more tightly. This would pose disturbing implications for the trend of energy and technology policy decision-making and — most important — for democratic process.

This is the far greater danger. The nation must confront directly the question of whether it is prepared to accept a chipping away at democracy as the price it will pay for nuclear energy. Disregarding or forcing out the public is not the answer. The alternative is instead to open up both the regulatory process and the terms of the discussion to address substantively the public's broader concerns — and then see what happens. This is not an industry that can be resurrected solely by the claims of scientific and technical experts, nor should it be. The determining questions about nuclear power are rooted instead in complex public policy and social concerns that must be resolved in open debate in the regulatory, social, and political arenas to the public's satisfaction. The far more important question is whether in today's political climate that debate will be encouraged or suppressed.

References

AEC (U.S. Atomic Energy Commission). 1950. *Summary Report of Reactor Safeguards Committee,* WASH-3. Washington, D.C.: U.S. Atomic Energy Commission.

Asselstine, James. 1987. "The Future of Nuclear Power After Chernobyl." *Virginia Journal of Natural Resources Law.* Volume 6, No.2 (Spring): 239-261.

Balogh, Brian. 1991. *Chain Reaction: Expert Debate and Public Participation in American Commercial Nuclear Power, 1945-1975.* Cambridge, MA: Cambridge University Press.

Bupp, Irvin C. and Jean-Claude Derian. 1981. *The Failed Promise of Nuclear Power.* New York, NY: Basic Books, Inc.

Cottrell, William B. 1974. "The ECCS Rule-Making Hearing." *Nuclear Safety.* Volume 15 (January-February): 30-55.

Clarfield, Gerald H. And William N. Wiecek. 1984. *Nuclear America: Military and Civilian Nuclear Power in the United States, 1940-1980.* New York, NY: Harper & Row.

Del Sesto, Steven L. 1980. "Conflicting Ideologies of Nuclear Power: Congressional Testimony on Nuclear Reactor Safety." *Public Policy.* Volume 28 (Winter): 40-70.

_____. 1979. *Science, Politics, and Controversy: Civilian Nuclear Power in the United States, 1946-1974.* Boulder, CO: Westview Press.

DOE. 1987. *Energy Security: A Report to the President of the United States.* Washington, D.C.: U.S. GPO.

DOE/EIA (U.S. Department of Energy, Energy Information Administration). 1993. *Annual Energy Review 1992.* Washington, D.C.: U.S. Department of Energy, Energy Information Administration.

_____. 1982. *U.S. Commercial Nuclear Power.* Washington, D.C.: U.S. Department of Energy, Energy Information Administration.

Farhar, Barbara C., Charles T. Unseld, Rebecca Vories, and Robin Crews. 1980. "Public Opinion About Energy." Pp. 141-172 in Jack M. Hollander, Melvin K. Simmons, and David O. Wood (eds). *Annual Review of Energy 5.* Palo Alto, CA: Annual Reviews Inc.

Ford, Daniel. 1984. *The Cult of the Atom.* New York, NY: Touchstone (Simon and Schuster).

Freudenburg, William R. and Eugene A. Rosa. 1984. "Are The Masses Critical?" In William Freudenberg and Eugene Rosa (eds). *Public Reactions to Nuclear Power.* Boulder, CO: Westview Press: 331-348.

GAO (U.S. General Accounting Office). 1989. *What Can Be Done to Revive the Nuclear Option?* GAO/RCED-89-67. Washington, D.C.: U.S. GAO.

_____. 1987. *A Perspective on Liability Protection for a Nuclear Plant Accident.*   GAO/RCED-87-124. Washington, D.C.: U.S. GAO.

_____. 1986. *Financial Consequences of a Nuclear Power Plant Accident.* GAO/RCED-86-193BR. Washington, D.C.: U.S. GAO.

_____. 1985. *Probabilistic Risk Assessment: An Emerging Aid To Nuclear Power Plant Safety.* GAO/RCED-85-11. Washington, D.C.: U.S. GAO.

_____. 1980. *The Nuclear Regulatory Commission: More Aggressive Leadership Needed.* EMD-80-17. Washington, DC: GAO

Gillette, Robert. 1972a. "Nuclear Safety (I): The Roots of Dissent." *Science.* Volume 177 (September 1): 771-776.

_____. 1972b. "Nuclear Safety (II): Years of Delay." *Science. Volume* 177 (September 8): 867-871.

_____. 1972c. "Nuclear Safety (III): Critics Charge Conflicts of Interest." *Science.* Volume 177 (September 15): 970-975.

_____. 1972d. "Nuclear Safety (IV): Barriers to Communication." *Science.* Volume 177 (September 22): 1080-1082.

Glitzenstein, Eric. 1994a. "The Role of the Public in the Licensing of Nuclear Power Plants." In David P. O'Very, Christoph E. Paine, and Dan W. Reicher (eds). *Controlling the Atom in the 21st Century.* Natural Resources Defense Council. Boulder, CO: Westview Press: 155-191.

_____. 1994b. "Public Participation in the Oversight of Nuclear Power Plant Operations." In David O'Very, Christopher Paine, and Dan W. Reicher (eds). *Controlling the Atom in the 21st Century*. Natural Resources Defense Council. Boulder, CO: Westview Press: 193-227.

Gordon, Joshua and Mark Knapp. 1989. *Consequences of a Nuclear Accident*. Washington, D.C.: Public Citizen.

Green, Harold, 1981. "Implications of Nuclear Accident Preparedness For Broader Nuclear Policy." In John W. Lathrop (ed). *Planning for Rare Events: Nuclear Accident Preparedness and Management*. Proceedings of an International Workshop. January 28-31, 1980. Oxford, UK: Pergamon Press: 157-166.

Green, Harold P. and Alan Rosenthal. 1963. *Government of the Atom*. New York, NY: Atherton Press.

Marrett, Cora Bagley. 1984. "Public Concerns About Nuclear Power and Science." In William Freudenberg and Eugene Rosa (eds). *Public Reactions to Nuclear Power*. Boulder, CO: Westview Press: 307-330.

Marshall, Eliot. 1983. "The Salem Case: A Failure of Nuclear Logic." *Science*. Volume 220 (April 15): 280-282.

Mazuzan, George T. and Roger R. Trask. 1979. *An Outline History of Nuclear Regulation and Licensing, 1946-1979*. Unpublished manuscript, Historical Office, Office of the Secretary, U.S. Nuclear Regulatory Commission. Washington, D.C.: U.S. Nuclear Regulatory Commission.

Mazuzan, George T. and J. Samuel Walker. 1985. *Controlling The Atom: The Beginnings of Nuclear Regulation 1946-1962*. Berkeley, CA: University of California Press.

Morone, Joseph G. and Edward J. Woodhouse. 1989. *The Demise of Nuclear Energy? Lessons for Democratic Control of Technology*. New Haven, CT: Yale University Press.

National Research Council. 1985. *Revitalizing Nuclear Safety Research*. Committee on Nuclear Safety Research,

Commission on Physical Sciences, Mathematics, and Resources. Washington, DC: National Academy Press.

Nealey, Stanley M. 1990. *Nuclear Power Development.* Columbus, OH: Battelle Press (Battelle Memorial Institute).

Nicodemus, Charles. 1994. "Lab Chief Admits Guilt On Testing For A-Plants." *Chicago Sun-Times.* (April 20): 22.

NRC (U.S. Nuclear Regulatory Commission). 1995. *Accident Source Terms for Light-Water Nuclear Power Plants.* L. Soffer et al, Office of Nuclear Regulatory Research. NUREG-1465. Washington, DC: U.S. GPO.

_____. 1994. *Annual Report 1993.* NUREG-1145, Vol. 10. Washington, DC: U.S. GPO.

_____. 1990a. *Severe Accident Risks: An Assessment for Five U.S. Nuclear Power Plants. Final Summary Report.* Division of Systems Research, Office of Nuclear Regulatory Research. NUREG-1150, Vol. 1. Washington, DC: U.S. GPO.

_____. 1990b. Memorandum from Samuel Chilk, Secretary, to James M. Taylor, Executive Director for Operations, June 15, 1990, Subject: SECY.-89-102 - Implementation of the Safety Goals.

_____. 1988. Generic Letter No. 88-20, November 23, 1988, and appendices.

_____. 1986a. *Reassessment of the Technical Bases for Estimating Source Terms.* NUREG-0956. Washington, DC: U.S. GPO.

_____. 1986b. Memorandum from Victor Stello, Executive Director for Operations, to Commissioners, May 19, 1986, Subject: Frequency of Severe Core Damage Accident.

_____. 1982. *Technical Guidance for Siting Criteria Development.* NUREG/CR-2239. David Aldrich et al, Sandia National Laboratories. Washington, DC: U.S. GPO.

_____. 1979. Press release and attached statement dated January 19, 1979.

_____. 1978. *Risk Assessment Review Group Report to the U.S. Nuclear Regulatory Commission.* Ad Hoc Review

Group, H.W. Lewis, Chairman.   NUREG/CR-0400. Washington, DC: U.S. GPO.

_____. 1975a. *Reactor Safety Study: An Assessment of Accident Risks in U.S. Commercial Nuclear Power Plants.* WASH-1400 (NUREG-75/014). Washington, DC: U.S. GPO.

_____. 1975b. *Reactor Safety Study: An Assessment of Accident Risks in U.S. Commercial Nuclear Power Plants, Executive Summary.* WASH-1400 (NUREG-75/014). Washington, DC: U.S. GPO.

Nuclear Energy Policy Study Group. 1977. *Nuclear Power Issues and Choices. Report of the Nuclear Energy Policy Study Group.* Cambridge, MA: Ballinger Publishing Company.

Okrent, David. 1981. *Nuclear Reactor Safety: On the History of the Regulatory Process.* Madison, WI: University of Wisconsin Press.

_____. 1987. "The Safety Goals of the U.S. Nuclear Regulatory Commission." *Science.* Volume 236 (April 17): 296-300.

Openshaw, Stan. 1986. *Nuclear Power: Siting and Safety.* London, UK: Routledge & Kegan Paul.

OTA (U.S. Office of Technology Assessment). 1984. *Nuclear Power in an Age of Uncertainty.* Washington, DC: U.S. GPO.

Perrow, Charles. 1984. *Normal Accidents: Living With High-Risk Technologies.* New York, NY: Basic Books.

President's Commission on the Accident at Three Mile Island. 1979a. *Report of the President's Commission on the Accident at Three Mile Island, The Need for Change: The Legacy of TMI.* Washington, DC: U.S. GPO.

_____. 1979b. *Staff Report to the President's Commission on The Accident at Three Mile Island: The Nuclear Regulatory Commission.* Washington, DC: U.S. GPO.

Rees, Joseph. 1994. *Hostages of Each Other: The Transformation of Nuclear Safety Since Three Mile Island.* Chicago, IL: University of Chicago Press.

Riccio, Jim. 1994. *Abuses of Discretion: NRC's Non-Enforcement Policy.* Critical Mass Energy Project. Washington, DC: Public Citizen.

Riccio, Jim and Matthew Freedman. 1993a. *Nuclear Lemons: An Assessment of America's Worst Commercial Nuclear Power Plants.* Critical Mass Energy Project. Washington, DC: Public Citizen.

_____. 1993b. *Hear No Evil, Speak No Evil: What the NRC Won't Tell You About America's Nuclear Reactors.* Critical Mass Energy Project. Washington, DC: Public Citizen.

Rogovin, Mitchell. 1980. *Three Mile Island: A Report to the Commissioners and the Public.* Nuclear Regulatory Commission, Special Inquiry Group, Mitchell Rogovin, Director, Washington, DC: U.S. G.P.O.

Rolph, Elizabeth S. 1979. *Nuclear Power and the Public Safety.* Lexington, MA: Lexington Books (D.C. Heath & Company).

_____. 1977. *Regulation of Nuclear Power: The Case of the Light Water Reactor.* Report No. R-2104-NSF. Santa Monica, CA: Rand Corporation.

Rosa, Eugene A. and Riley E. Dunlap. 1994. "Nuclear Power: Three Decades of Public Opinion." *Public Opinion Quarterly.* Volume 58 (Summer): 295-325.

Sholly, Steven. 1995. Unpublished analysis of compiled data from Individual Plant Examinations (IPE). Personal communication, January, 1995.

Sholly, Steven C. and Gordon Thompson. 1986. *The Source Term Debate: A Report by the Union of Concerned Scientists.* Cambridge, MA: Union of Concerned Scientists.

Simpson, John. 1994. "Nuke Groups at Odds Over Safety Standards." *Public Utilities Fortnightly.* (February 1): 37-38.

Slovic, Paul. 1993. "Perceived Risk, Trust, and Democracy." *Risk Analysis* 13: 675-681.

Slovic, Paul. 1991. "Perception of Risk and the Future of Nuclear Power." In M. Golay (ed). *Proceedings of the First MIT International Conference on the Next Generation of Nuclear Power Technology.* Cambridge, MA: Massachusetts Institute of Technology, Program for Advanced Nuclear Studies (reprint obtained from author, 11 pp.): Section 6, 1-8.

Slovic, Paul, Baruch Fischoff, and Sarah Lichtenstein. 1982. "Rating the Risks: The Structure of Expert and Lay Perceptions." In Christoph Hohenemser and Jeanne X. Kasperson (eds). *Risk in the Technological Society.* Boulder, CO: Westview Press: 141-166.

Smith, Kirk R. 1988. "Perceptions of Risk Associated With Nuclear Power." *Energy Environment Monitor.* Volume 4, No.2: 61-70.

UCS (Union of Concerned Scientists). 1985. *Safety Second: A Critical Evaluation of the NRC's First Decade.* Cambridge, MA: Union of Concerned Scientists.

_____. 1977. *The Risks of Nuclear Power Reactors.* Cambridge, MA: Union of Concerned Scientists.

U.S. Congress, House. U.S. Congress, House. 1987. Subcommittee on General Oversight and Investigations, Committee on Interior and Insular Affairs. *NRC Coziness With Industry: An Investigative Report.* 100th Congress, 1st Session. December, 1987.

_____. 1986a. Committee on Energy and Commerce, Subcommittee on Energy Conservation and Power. *Hearings on Nuclear Reactor Safety.* Serial No. 99-177.99th Congress, 2nd Session. May 22 and July 16, 1986.

_____. 1986b. Committee on Energy and Commerce, Subcommittee on Energy Conservation and Power. *Hearing on Emergency Planning at Seabrook Nuclear Powerplant.* Serial No. 99-180. 99th Congress. 2nd Session. November 18.

_____. 1985. Committee on Energy and Commerce, Subcommittee on Energy Conservation and Power. *Hearing on NRC Authorization for Fiscal Years 1986-87.* Serial No. 99-22. 99th Congress, 1st Session. April 17, 1985.

Utroska, Daniel. 1987. "Holes in the U.S. Nuclear Safety Net." *Bulletin of the Atomic Scientists.* (July-August): 36-40.

van der Plight, Joop. 1992. *Nuclear Energy and the Public.* Oxford, UK: Blackwell Publishers.

Wald, Matthew. 1993. "Nuclear Plant Operators are Told of a Safety Risk." *New York Times* (June 1): A8.

_____. 1989. "Nuclear Safety Goals Are Not Met." *New York Times* (March 27): D4.

Walker, J. Samuel. 1992. *Containing the Atom: Nuclear Regulation in a Changing Environment 1963-1971.* Berkeley, CA: University of California Press.

Chapter 6

Waste Disposal and Decommissioning

Carolyn Raffensperger

This abstract, hollow junk seems
beautiful and necessary. . .
prize your flaws, defects,
behold your accidents,
engage your negative criticisms,
these are the materials of your ongoing.

(Ammons, 1993)

Introduction

The nuclear power industry is creeping past middle age,
and not doing it gracefully. As the industry ages, new problems
surface. This is most evident as the industry begins planning
decommissioning of fuel cycle facilities. While radioactive waste
has always been the flaw in the system of presumably clean, safe
energy, few considered the enormity of the waste problem when
nuclear power plants reached the end of their useful life.[1] And the
major issue in decommissioning is how much waste is generated

---

[1] In fact, the International Atomic Energy Agency did not hold its first meeting
on decommissioning until 1973, almost 20 years after the first reactor was built
(Pollack, 1986: 40).

and where it will go. It has been said that the "essential challenge of decommissioning is to remove and dispose of radioactive waste, while keeping occupational and other exposures as low as possible" (Office of Technology Assessment, 1993: 108). This challenge is far from being met, despite the maturity of the industry.

Three interlocking political realities shape the debate around decommissioning and decontaminating (D&D) nuclear power facilities: 1) the fact that radioactive waste disposal has reached a standstill; 2) federal agencies, particularly the Nuclear Regulatory Commission,[2] have proposed innovative new regulations which will radically change D&D, and; 3) ongoing tensions between scientists, the public, national and local decision makers regarding site-specific nuclear issues.

Decommissioning is defined by the Nuclear Regulatory Commission as removing "nuclear facilities safely from service and reducing] residual radioactivity to a level that permits unrestricted use and termination of the license. Decommissioning activities are initiated when a licensee decides to terminate licensed activities." Decontamination is the piece of decommissioning which reduces or eliminates radioactive contamination of the site and equipment (Thompson and Goo, 1994: 32-41).

D&D is undertaken to protect the public and the natural environment from accidental releases of radioactivity (Office of Technology Assessment, 1993:110). It is important to make explicit that decommissioning does not eliminate contaminants but

---

[2] The Nuclear Regulatory Commission (NRC) has primary jurisdiction over the D & D of all non-federal facilities. The Environmental Protection Agency (EPA) has jurisdiction over Superfund sites which may contain radionuclides. EPA also has jurisdiction over federal fuel cycle facilities, including the uranium enrichment facilities. Some states have been delegated jurisdiction as agreement states. For D & D, the NRC plans on requiring agreement states to have identical regulations.

only isolates and transfers them from one site to another (Office of Technology Assessment, 1993: 107). The process of D & D moves contaminants from the fuel cycle facility to treatment, storage, and disposal facilities. This is important in any discussion about risk because it acknowledges that the hazards can only be contained, not eliminated.

The NRC estimates the population of nuclear fuel cycle facilities[3] which will require D&D to include 112 nuclear power plants at 75 sites; 74 research and test reactors, 14 fuel fabrication plants, 2 uranium hexafluoride production plants, 49 uranium mill facilities and 9 independent spent fuel storage installations. In addition, the Environmental Protection Agency (EPA) will oversee D & D of the three uranium enrichment facilities which have been managed by the U.S. Department of Energy (DOE).

Some commentators, both in industry and in the regulatory arena, have argued that the process of D&D at many non-fuel cycle facilities can be conducted as a standard industrial demolition project. Thus, the Office of Technology Assessment, has argued that "although no large commercial reactors have undergone complete decommissioning . . . the task of decommissioning [such] plants can be accomplished with existing technologies. . . Many of the conventional technologies used to decommission nuclear power plants are the same ones used to demolish other industrial facilities and buildings" (1993: 30). For at least two reasons, however, such an assessment seems profoundly unwarranted. First, in addition to the large amount of radioactive waste generated during normal operations, large areas of the facility as well as the host site

---

[3]   Fuel cycle facilities include power, test, and research reactors, uranium fuel fabrication plants, uranium hexafluoride conversion facilities, uranium mills, and independent spent fuel storage installations (NUREG 1496 Vol. 1:1994: 3-8). These are the "facilities involved in any of the steps leading to or resulting from the generation of electricity by controlled nuclear fission of uranium" (NUREG 1496 Vol. 1:1994: 4-2).

have been contaminated with radionuclides. While the problem of disposal of this contaminated material has produced much political debate, it has not, to date, produced a solution as to how and where to store the waste. Thus, while there are ample storage sites to dispose of conventional demolition debris, there are currently no facilities to take the waste resulting from D&D and, most importantly, no one can predict when such facilities will be operational. Nor will we know how much waste will be generated as we shut down the aging power plants until decision makers assess the possibilities of future site use and recycling contaminated materials.

A second reason for not conceiving of D&D as a standard demolition project concerns the problem of cost. While D&D will cost significant sums of money, little useful can be said about that cost because projections are so speculative and uncertain (Pollack, 1986: 25). In large part, this is due to the limited experience the United States has in decommissioning nuclear facilities, a fact even the optimists readily concede. The uncertainties include the extent of site contamination at closure, labor and waste disposal costs, and regulatory standards at the time of D&D (Office of Technology Assessment, 1993: 132). But low-level radioactive waste disposal is estimated to be over one-third of the total cost for a nuclear power plant which is immediately dismantled (Office of Technology Assessment, 1993: 108).

These issues look like national issues on the surface. And certainly with Congress, federal agencies and the Supreme Court involved, they are national issues. Clearly, however, the impacts of these decisions are most clearly expressed at the local level. Will low-level radioactive waste disposal facilities be located in small and relatively powerless communities? Will the nuclear power plant in local communities be cleaned up to the extent that the land can be used for another purpose? The interplay between local and national decision making can be contentious, pitting the interests of each level against the other. For example, local environmentalists may not agree with their national counterparts;

nor may local communities see the needs and benefits of waste disposal in the same way that national constituencies might.

This chapter focuses on radioactive waste disposal and national policies which govern D&D, particularly the NRC's proposed rule.   It addresses the questions of nuclear facility decommissioning. Waste disposal siting, and facility safety. And it raises the core concern: who should decide issues surrounding these matters?

Waste Disposal

There are three kinds of radioactive waste which are proving to be an intractable problem for the nation: low-level radioactive waste, high-level radioactive waste and mixed waste.[4] These wastes, particularly low-level and mixed wastes, are generated through normal operations of many kinds of industries from medicine to nuclear power plants. During the process of D&D of the nuclear fuel cycle facilities, only low-level radioactive waste is a major issue. There are no national projections of how much mixed waste will be generated during D&D of nuclear facilities. Spent fuel, that is, high-level waste, must be disposed of as part of the operating license and is, therefore, not part of D&D. However, the absence of a high-level waste repository may significantly delay license-termination and D&D of all commercial nuclear power plants.

D&D is certainly the wild card in the radioactive waste poker game. Operating facilities generate predictable quantities of waste. But, the volume of waste generated through decontamination and decommissioning is uncertain due to three considerations: 1) the point in time when a facility will be decontaminated after its operating life is complete; 2) final disposition of contaminated materials; and 3) whether the facility

---

[4]   Mixed Waste is waste that contains both a chemical and radioactive hazard.

will be released for unrestricted use or will remain under some form of institutional control and restriction.

Timing of Decontamination and Decommissioning: Decon, Safestor, Entomb

The NRC encourages, but does not require, completion of decommissioning within 60 years of plant closure.[5] Within that general time frame, three scenarios are generally considered for scheduling the final decommissioning of nuclear power plants at the end of their operating life. They can either be immediately dismantled, put into safe storage with later dismantlement, or be permanently entombed. (The League of Women Voters Education Fund, 1993: 68). The NRC has respectively labeled these three options *Decon, Safestor* and *Entomb*.

Decon would generate the most waste to be shipped off-site because the contaminating radionuclides would not have decayed to insignificant levels and would have to be disposed of at a low-level radioactive waste facility. Two nuclear power facilities have been decommissioned and three are in process. The two decommissioned facilities are Elk River in Minnesota, and Shippingport in Pennsylvania. Fort St. Vrain, Saxton and Shoreham are being decommissioned now (The League of Women Voters Education Fund, 1993: 68). Shippingport was an important lesson for those interested in D&D because it was the most comprehensive D&D experience to date. It was dismantled in 4 years and at a cost of $91.3 million dollars (Office of Technology Assessment, 1993: 112). However, numerous factors make it difficult to extrapolate from Shippingport to larger facilities that are privately owned. For instance, if the D&D waste from Shippingport had gone to Barnwell, the only commercial facility then available, instead of a DOE facility, the project cost would

---

[5]   10 CFR 30.4:50-2 and 50.8.

have been an additional $56 million dollars over the $91.3 million dollar price tag (Office of Technology Assessment, 1993: 108).

The second scenario, Safestor, entails closing the plant and securing it from public intrusion to permit decay of radioactivity (The League of Women Voters Education Fund, 1993: 70). The plant then undergoes D&D. Safestor generates less radioactive waste when it is finally decommissioned because of the decay of contaminating radionuclides. Waiting 50 years to decontaminate a power plant could reduce low-level radioactive waste by up to 90 percent because radioactivity declines dramatically during the first 30 years of storage (Office of Technology Assessment, 1993: 108; Pollack, 1986: 25).

Industry has some experience with Safestor. An experimental reactor in Santa Susana California was safe-stored in 1964. Actual D&D began in 1974 and was completed in 1983. There are also a number of commercial reactors currently in Safestor including LaCrosse in Wisconsin, and Rancho Seco in California (The League of Women Voters Education Fund, 1993: 70). However, there are public relations problems with aging power plants sitting idle in communities. Rusting, paint-peeling structures do not make good neighbors; not to mention the dangers and risks posed by a decommissioned plant (from radiation to terrorists attack).

Entombment is permitted for facilities that are contaminated with short-lived radionuclides (The League of Women Voters Education Fund, 1993: 70). The site and facility are decontaminated as feasible, the reactor is sealed and the site is placed on restricted access until the radioactivity has decayed to unrestricted release levels. This scenario generates the least amount of waste but requires institutional controls for as long as 100 years. Institutional controls are notoriously difficult to maintain over long time spans. Three small experimental reactors have been entombed — Hallam (Nebraska), Piqua (Ohio), and Ricon, Puerto Rico.

Disposition of Contaminated Materials

There are no commercial facilities accepting low-level radioactive waste at this point in time. And it is unlikely that any disposal facilities for these wastes will be operational for years to come. While this poses immediate problems for research institutions and small manufacturers of materials such as radiopharmaceuticals, the nuclear fuel cycle facilities have on-site storage capacity for operating waste. However, when the fuel cycle facilities cease operations and go through the process of D&D, waste becomes the single largest economic and environmental problem. In fact, decommissioning a large commercial power plant could generate more waste than the plant generated during its operating life.

The NRC currently estimates the volume of waste from D&D of existing land and structures as 15 million cubic feet (NUREG 1496 Vol. 1,1994: 5-27). This does not include D&D waste from uranium enrichment facilities. Based on an analysis using a 1,095 megawatt PWR reactor as a reference facility, D&D of nuclear power plants alone is estimated to involve 1.1 million cubic feet (NUREG 1496 Vol. 2, 1994: Appendix G). The NRC estimates that its reference reactor would have contaminated 250,000 square feet of flooring and 300, 000 square feet of walls. In addition, 5,300 square feet of soil around the reactor is predicted to be contaminated.

What will be done with this large and dangerous volume of waste? There are three possibilities: 1) storage of selected materials in a low level radioactive waste facility; 2) materials recycling; and 3) permanent or temporary in situ storage as suggested in the Safestor and Entombment scenarios. It is likely that all three possibilities will be employed at the nuclear fuel cycle facilities.

Low-Level Radioactive Waste Facilities

To this point, public effort has concentrated mostly on siting low-level radioactive waste facilities. Unfortunately, siting low-level waste disposal facilities has reached a logjam.

Before 1979, the LLRW disposal situation was straightforward and non-controversial. There were six commercial facilities which had been licensed in the 1960s and 1970s. By 1978 there were only three facilities open. This was described as a "happy situation for 47 states" because only three states had to be concerned with disposal and siting (Brown, 1989: 101). But in 1979, due to a series of accidents, two of the remaining sites were closed and industry began questioning whether, in the long run, there would exist sufficient storage and disposal capacity for the waste being generated.

In 1979 as a response to the crisis, the U.S. Congress searched for a federal solution and even proposed siting a national facility (U.S. Congress, 1985: 3007). However, the states asked for a delay and the following year Congress adopted a "state solution" recommended by the National Governor's Association and the National Conference of State Legislatures. Following this delay Congress passed the Low-Level Radioactive Waste Policy Act in 1980 (LLRW Act). The LLRW Act did several things, most notably making each state responsible for waste generated within its borders. But it also made clear that the disposal of low-level radioactive waste can be most safely and effectively managed on a regional basis.

After passage of the LLRW Act, the states struggled with economic and environmental solutions to the radioactive waste problem. There were two key issues. First, it seemed equitable to most states that whoever generated the waste should be responsible for storing it. Thus it was logical that large generators of waste, like Illinois and California, would host facilities. Needless to say, states that generated very little waste were not interested in hosting such a problematic facility.

Second, most states wanted to avoid the serious environmental problems experienced by past facilities. All previous facilities had used shallow land burial techniques which meant little more than throwing containers in the ground and bulldozing dirt over them. Illinois, New York and Kentucky had experience with leaking low-level radioactive waste facilities. Sheffield, Illinois was open from 1967-1978 and took approximately 3 million cubic feet of waste. In 1976, tritium was detected in the ground water in a nearby lake and in public and private water supplies (Illinois Department of Nuclear Safety, 1987: 15). Maxey Flats in Kentucky and New York's West Valley site had similar histories of burial and leakage off-site.

The Act was amended in 1985 by a Congress frustrated with the delay in siting new facilities. The Amendments set up a timetable that the compacts and go-it-alone states had to meet. There were exacting penalties and rewards attached to the milestones in the timetable. The amendments established new procedures and deadlines for development of new disposal facilities, surcharges to help finance those facilities, and allocations for utility disposal of waste at existing sites (Illinois Department of Nuclear Safety, 1990: 118). It also extended out-of-region access to existing disposal facilities to give regional compacts more time to site facilities (U.S. Department of Energy, 1986: 2).[6]

---

[6]   The Act's amendments were challenged by New York in a case that went before the U.S. Supreme Court. New York won its lawsuit and that section of the law which required states to take title to waste generated in their jurisdictions if the state did not site a facility by 1993 was overturned by the Court.

Despite this tortuous history and the expenditure of millions of dollars, not a single facility has been sited in the ensuing fourteen years.[7] The last remaining commercial site at Barnwell South Carolina ceased taking waste in July 1994. Only Richland in Washington state, an enormous DOE complex, is taking waste, (mostly) from the Northwest and Rocky mountain states, as well as DOE facilities. This leaves most of the country, particularly large generating states like California and Illinois, without any place to dispose of its waste.

Several factors coalesced to make siting low-level waste disposal facilities unusually difficult. In theory, the facilities looked good to states because the federal government had done a bad job of managing nuclear issues. However, when individual sites were chosen, local communities felt it onerous to take waste which was rarely generated in their community. As such, siting radioactive waste facilities became a matter of environmental justice, much like siting solid waste and hazardous waste incinerators and landfills. It appeared to many observers that poor communities were being chosen to host facilities and then bribed by the promise of jobs and money to fund local government activities. This often divided potential host communities.

There have also been technical and policy problems with low-level waste disposal, including the widely recognized and troubling problem of the definition of low-level waste. The LLWR Act defines the term low-level radioactive waste as radioactive material that (a) is not high-level radioactive waste, spent nuclear fuel, or byproduct material (as defined in section 2014 (e)(2) of this title) or (b) material that the Nuclear Regulatory Commission,

---

[7] Illinois spent over $80 million considering Martinsville as a host site. That site was unanimously rejected by a three-person Commission (of which I was one member). The state will be looking for a new site and begin the entire process again.

consistent with existing law and in accordance with (a), classifies as low-level radioactive waste.

This is an odd definition because the Act defines low-level radioactive waste by what it is not and/or as what NRC determines it to be. Neither definition offers much guidance to the states or the industry. The definition of low-level waste is somewhat clarified in federal regulation 10 CFR 61.55 which divides it into four categories; Class A, B, C and greater than Class C. Wastes fall into one or another category based two factors: (a) the degree to which the material must be segregated from various kinds of waste to insure its safe handling and the safety of the workers on site; this, in turn, is based upon the nature and type of radionuclides found in the waste; and (b) the steps which must be taken to ensure stability, i.e., assurance that the waste does not structurally degrade and affect overall stability of the site through slumping, collapse or other failure of the disposal unit and thereby lead to water infiltration (10 CFR 61.55 and 51.56). For the most part, Class A waste is dry active waste. It generally contains low-levels of radionuclides with very short half-lives. Class A will provide the greatest volume of waste in D&D of nuclear power plants. Both Class B and C waste include extremely hazardous material, most of which is generated by nuclear power plants. Class B and C can also contain materials with radionuclides that have very long half-lives and are mobile in water. In fact, some of the waste is liquid and, thus, subject to spills and accidents.

Volume Reduction and Recycling

Volume reduction has been identified as a possible solution to the low-level radioactive waste disposal problem. If waste volume can be reduced, facilities can be smaller in size, less material needs to be transported, and industry costs can be contained. Waste volume can be reduced in any number of ways. One of the most common techniques is treating the waste and then only having to dispose of the residue. Treatment techniques include scabbling, sandblasting, nuclear laundries, and compaction. One of the most controversial volume reduction techniques is recycling.

Of the projected D&D materials which will need treatment and disposal, soils, concrete and metals are the largest categories by volume. Out of those three, metals are the most likely candidates for large-scale recycling.   Current projections of radioactive scrap metal generated by decommissioning nuclear power plants prior to the year 2000 are 93,000 tons of steel and stainless steel, and 2,800 tons of copper (Clemens, 1993). The U.S. Department of Energy currently has 1.3 million tons of steel and stainless steel and 38,000 tons of copper stockpiled. The value of this metal estimated in 1990-1991 dollars is somewhere between $6.6 million and $130 million (Clemens, 1993).   These projections do not include the quantities of metals that will be generated from D&D of federal facilities, including uranium enrichment facilities.

From an environmental and public health perspective, recycling metals is a double edged sword: while recycling reduces the volume of contaminated materials that must be disposed, there is a substantial risk that recycling could increase the potential exposure to the public.   Pollution prevention and waste volume reduction have been the watchwords of the environmental movement for a decade.   And recycling irradiated scrap metal appears to meet those goals but at the cost of increasing exposure to the public.   Recycling contaminated metal appears to put environmental concerns in conflict with protection of human health.

Recycling irradiated metals can take one of two paths: (a) it can be restricted in its use and only allowed to be used within the nuclear industry; or (b) it can be released into the commercial scrap metal market. Recycling is particularly attractive if the waste can be reused within the nuclear industry in a manner which prevents clean metals from becoming contaminated. For instance, DOE has suggested that waste canisters for radioactive waste be manufactured out of contaminated scrap metal and then disposed of in a radioactive waste facility which still isolates the hazard from the public (Lilly, 1993).

The most contentious issue, however, is whether there should be a standard for free release of contaminated scrap metal into the commercial stream.   The difficulties of recycling contaminated metals and the potential for inadvertent mistakes yielding unacceptable contamination in the commercial stream have been demonstrated in 250 cases reported in Canada and the United States.   These cases include the discovery of metal smelts containing radioactivity. Most of the contamination is either cobalt 60 or cesium 137, radium 226, or naturally occurring radioactive materials (Yusko, 1993).   It is not unreasonable to presume that adopting a "free-release" standard will lead to widespread increased contamination of the commercial scrap metal supply either through inadvertence, intentionally unloading contaminated metals to avoid the cost of disposal, or concentrating radionuclide contamination through the recycling process.

The U.S. EPA is currently proposing a rule on recycling scrap metal. There may be wisdom in using irradiated metals within the DOE complex and other parts of the nuclear industry. It is also possible that protection of both human health and the environment can be met by recycling metals within the nuclear industry. However, a rule which permits free release of LLRW materials into the commercial waste stream raises a host of problems.   The dangers in such a policy direction were demonstrated in a prior NRC attempt to administratively downgrade waste.

In 1990 the NRC moved to deregulate radioactive materials which were what the NRC called "Below Regulatory Concern" (BRC), that is, materials that would offer exposure of 10 mrem per year (55 Federal Register 27522). It was proposed that materials falling below BRC level could be disposed of in municipal sanitary landfills or incinerators rather than expensive low-level radioactive waste disposal facilities. This would have represented major savings to industry, particularly nuclear power plants that generate the bulk of such waste.

But the proposed BRC rule created strange bedfellows in opposition to it. It was predictable that environmental, and antinuclear groups would oppose the rule which permitted increased doses of radiation to an unsuspecting public.[8] What was not predicted was that compact host states who had to site low-level waste facilities would also oppose the BRC rule. Their opposition stemmed from the possibility that the rule would undermine the economics of the LLRW facilities by removing a large volume of waste upon fee structures were based. Given the possibility that there could be 10 new disposal facilities, economies of scale required that all the projected waste go to the facility slated for each compact. The BRC rule seriously challenged that economic viability. The state-level response to the NRC proposal was immediate and unambiguous: press conferences were held, officials testified before the NRC, and citizens were organized, all speaking out against administrative downgrading (Kraft, 1990).

The opposition was so fierce that the NRC was forced to withdraw the rule. Congress went on to revoke the NRC's policy on BRC in the 1992 Energy Policy Act. This was a stunning contretemps to the NRC.

---

[8]   In a press conference, Dr. Michael McCally of Physicians for Social Responsibility said the BRC proposal "could be likened to randomly firing a bullet down the streets of Manhattan, because the actual 'risk' to any single individual is [probabilistically] low. With an 'acceptable' standard for BRC of one fatal cancer per hundred thousand people per year, this will mean an additional 2,500 cancer fatalities per year from BRC." (Kraft, 1990).

NRC's Proposed Rule on
Decontamination and Decommissioning

In an attempt to regroup from the BRC debacle, the NRC retreated, and developed a new, and somewhat novel, rulemaking process. First, it selected a real problem — decommissioning and decontamination — rather than the mostly philosophical issue of BRC. The goal of the rulemaking was to provide specific radiological criteria for the decommissioning of lands and structures (59 Federal Register 43200).

Second, the NRC developed an enhanced participatory rulemaking process. The NRC held seven workshops around the country from January through May of 1993 to gain input into the rule design. The NRC attempted to involve all stakeholders, from environmentalists to health physicists and industry. The workshops were designed to explore issues, not achieve consensus. This open process was an unusual exercise for an agency born out of the long history of secrecy surrounding the atomic bomb.

Third, the NRC worked in collaboration with EPA in drafting the rule, signing a Memorandum of Understanding which provides the framework for that collaboration. EPA has jurisdiction over federal facilities such as the uranium enrichment facilities. But EPA also has statutory responsibility to establish standards protecting the public from radioactive material outside the NRC licensees site boundaries. EPA is currently drafting those standards. If EPA is satisfied with the NRC rule it will suspend applicability of its rule to NRC's licensees. Past interagency conflict has led to formidable impasses. For instance, different EPA and NRC regulations governing mixed waste prevented any solutions since land disposal, prohibited for hazardous waste, is the *only* option for radioactive waste (Office of Technology Assessment, 1993: 116).

While the process has been truly remarkable, the result still has controversial elements. The goal of the proposed rule is that decommissioning should reduce residual radioactivity at a site to levels that are indistinguishable from background, allowing the site to be available for unrestricted use and subsequent termination of the license (59 Federal Register 43205-43206). The site release limit is 15 mrem (0.15 mSv/year) Total Effective Dose Equivalent (TEDE) to an average member of the Critical Group for residual radioactivity distinguishable from background (59 Federal Register 43229). The TEDE is calculated for the first 1000 years following decommissioning.

One of the most artful puzzle pieces in the rule is a requirement that the licensee must show that residual radioactivity from the site will not cause radioactivity in groundwater to exceed EPA's drinking water standard in 40 CFR Part 141 (59 Federal Register 43229). This requirement demonstrates the value of interagency partnerships: it is ridiculous to have several standards applicable to radionuclides in water. But more importantly it reflects an assessment of which health hazards are worrisome. The TEDE is designed to avoid serious radiation exposure during within the first 1000 years after decommissioning. Groundwater, drinking water in particular, is the greatest concern over that time span.

However, the draft rule provides that, under some conditions, the Commission will terminate the license but release the site to restricted use and institutional control (59 Federal Register 43229). The TEDE from residual radioactivity may not exceed 100 mrem (1mSv) per year even if the restrictions were ineffective. A decommissioning plan for restricted release triggers a requirement that the licensee convene a Site Specific Advisory Board (SSAB) comprised of members of the public. The SSAB's purpose is to provide advice to the licensee on details surrounding the proposed decommissioning of the site.

The section of the rule on restricted use has raised the ire of both industry and the environmental community. Industry feels that public participation in the form of the SSAB could delay site cleanup, increase the cost, and require industry to cater to outrageous demands from what they perceive to be antinuclear extremists. On the other hand, the environmental community feels that the NRC is permitting industry to walk away from a dirty site with little liability.[9] It does not seem appropriate that a license can be terminated with contamination left at the facility. It appears from the rule that the site could be used for another industrial use, increasing exposure to workers in the new industry to higher doses of radiation than if that site had been cleaned up to background levels. This, environmentalists argue, is not consistent with the guiding principle of health physics which is that exposure is ALARA or, "as low as reasonably achievable" (59 Federal Register 43229).

Who Decides?

As Gerald Holton, a physicist at Harvard, has observed "the fates of science, technology, and society have become linked in ever more complex ways, each of the three being shaped as much by the other two as by its own dynamics" (Mathews, 1994a: 73). The history of the old BRC rule and NRC's proposed D&D rule point out that the public expects to have a role in decisions about the impact of science and technology on their lives, particularly when the technology is nuclear power. While national public interest groups expect to participate in the formation of policy, local groups expect to have a voice in site-specific decisions. Scientists, public interest groups and local citizenry all desire a role

---

[9]   It should be noted that the rule does require "sufficient financial assurance to enable an independent third party to assume and carry out responsibilities for any necessary control and maintenance of the site." Part of the SSAB's role is to examine this question (59 Federal Register 43229).

in these decisions which are characterized as having a high degree of scientific uncertainty and a large set of societal values at stake.

It would seem, however, that scientists and the public have reached an intellectual standoff. Many scientists believe that nuclear power plants and/or disposal sites pose little danger. The public remains unconvinced (see Greenberg, this volume). The tension between the scientist and the public in the nuclear arena points out the collision between specialized knowledge and public knowledge in a democratic society (Mathews, 1994b). Scientists bring a specialist's knowledge of which radionuclides are highly mobile in groundwater or whether an aqueous or gaseous cleanup technique is preferable. But the public brings an understanding of land values, how many truck loads of soil passing through the neighborhood are appropriate and whether an increased cancer risk of one in ten thousand is acceptable. Public knowledge is the expertise and ethics citizens hold concerning their locale. It is the wisdom of place and community. It is no less valuable than the specialized knowledge scientists possess.

The history of the rule on BRC and the disagreements over the NRC's decommissioning rule are prime examples of the disputes between science and the public. This tension is not a new story. But it leaves unanswered the question of how we proceed in the face of that tension.

Decision Making that Incorporates Societal Values and Scientific Knowledge

Decisions about large environmental and public health issues, with all the attendant uncertainty and complexity that accompany them do not lend themselves to either a standard scientific process or the usual political treatment. They will require a new approach. In their article entitled "Uncertainty, Complexity and Post-Normal Science," Funtowicz and Ravetz (1994) suggest that scientists need a new model of science-society interaction to address complex problems with uncertain impacts. They propose that scientific input into policymaking moves beyond formalized deduction to an interactive dialogue with stakeholders.

The issues that Funtowicz and Ravetz have in mind are characterized by irreducible complexities and uncertainties within the system and by a host of costs, benefits, and value commitments brought to the table by stakeholders. Funtowicz and Ravetz call these "system uncertainties" and "decision stakes" (1994: 1881). When the system uncertainties and decision stakes are high, the authors argue that decision makers are faced with post-normal science. Post normal science "indicate[s] that the puzzle-solving exercises of normal science . . . which were so successfully extended from the laboratory of core science to the conquest of nature through applied science, are no longer appropriate for the solution of . . . problems" (Funtowicz and Ravetz, 1994: 884)

In post-normal science, the uncertainties can be ethical or epistemological. Soft values have as much weight as hard facts. In fact, facts and values often cannot be separated in a meaningful way. Funtowicz and Ravetz suggest that the most effective way to solve problems in the face of severe uncertainties and high value conflicts is to engage in dialogue, tolerating the initial confused phases and ambiguity until new research can be stimulated. The goal of this process of consensus is to determine the kind of scientific framework in which research will be carried out.

Decontamination and decommissioning of nuclear facilities fall into the purview of post-normal science, given the level of uncertainty and complexity inherent in the decisions that most be made. Funtowicz and Ravetz pose a way that these decisions can be made: extensive dialogue between citizens and scientists at the initial stages of the cleanup process to determine what scientific framework will guide the process. In many ways this can ease the tension between scientists and the public, between specialized knowledge and public knowledge. It is unlikely that we will either site new waste facilities or adequately clean up existing nuclear facilities until scientists and citizens are brought into interactive dialogue about the values and science at stake.

National vs. Local Decision Making

The political landscape is changing as much as the scientific landscape. Over the last twenty years most policy decisions related to nuclear power were made through the federal legislative process. The environmental movement, in particular, has been ingenious at shaping the national agenda. Groups like the Sierra Club, the Audubon Society, the Natural Resources Defense Council and others have been instrumental in using national laws such as the Clean Air Act to intervene in nuclear issues. These laws have done much to change the way the nation lives and does business.   Oftentimes, however, the laws tend to ignore a fundamental fact that governs American politics: implementation must usually be done on the local level. This is certainly true for decommissioning and decontamination of nuclear facilities. Once the NRC finalizes its rule on D&D, actions are going to be site-specific, and so are impacts and risks.

Low-level radioactive waste has always been a regional or local issue and has raised disagreements between national groups and local citizens: it is just as difficult to reconcile values across national and local boundaries as it is to reconcile values and science between scientists and the public. Consensus breaks down when policy on national waste disposal must be implemented at a specific location.

Where public debate most impacts D&D policy is on the question of how much residual radioactivity should be left at a site. How clean is clean enough? National groups are fighting hard to have a rule that prohibits license termination with restricted use. However, it may well be that local communities will decide that hauling additional tens of thousands of truckloads of contaminated soil through their neighborhood is not worth the decreased risk of exposure to radiation.   How do we make wise political decisions on a national level and yet provide flexibility to local citizens to make the policy choices concerning implementation as these facilities are decommissioned? Finding creative ways of making these decisions on the national and local levels will, of necessity, engender new ways of protecting the environment and public health and safety.

Conclusion

As the U.S. considers decommissioning of its nuclear facilities, we are faced with difficult choices.   Decommissioning is expensive and involves complexities and uncertainties that are well beyond typical policy discourse. The retirement of a nuclear plant requires disposal of wastes generated during its operating life, as well as a quantity of waste from decommissioning itself that may exceed in volume the waste from plant operations.   It is difficult to work out wise public policy among scientists, the public and national and local decision makers as to where and how to dispose of this waste.   D&D policy represents an opportunity to engage in a debate on one of the most important issues of our generation and one that will significantly affect future generations. We must, for intergenerational equity, for the sake of the environment and for our own sanity, find a course that protects public health and the environment in a way that also honors the local wisdom of place and community.

References

Ammons, A.R. 1993. "Garbage." In Louise Gluck (ed). *The Best American Poetry of 1993.* New York, NY: Macmillan Books.

Brown, Holmes. 1989. "Low-level Waste Policy: Past History and Future Expectations." In *Nuclear Safety: Power*

Production, Waste Disposal Conference Proceedings. Springfield, Il: State of Illinois.

Clemens, Bruce. 1993. "Establishing Standards for Radioactive Scrap Metal." In *Proceedings of the Radioactive Scrap Metal Conference.* Knoxville, TN: Energy, Environment and Resources Center at the University of Tennessee.

Federal Register. 1994. Vol. 59 #161 *Proposed Rules: Nuclear Regulatory Commission 10 CFR Parts 20,30,40,50,51,70 and 72.*

Funtowicz, Silvio O. and Jerome R. Ravetz. 1994. "Uncertainty, Complexity and Post-Normal Science." *Environmental Toxicology and Chemistry.* Volume 13, No.12 (December): 1881-1885.

Illinois Department of Nuclear Safety. 1987. *Annual Report 1986-1987.* Springfield, Il: State of Illinois.

Illinois Department of Nuclear Safety. 1990. *1989 Annual Survey Report.* Springfield, Il: State of Illinois.

Kraft, David. 1990. "Illinois Groups: 'BRC' Not Acceptable." *NEIS News.* Volume 9, No.1 (Spring).

League of Women Voters Education Fund. 1993. *The Nuclear Waste Primer: A Handbook for Citizens.*

Lilly, Judson. 1993. "U.S. Department of Energy Involvement in Recycling and Resuse of Radioactive Scrap Metal." In *Proceedings of the Radioactive Scrap Metal Conference.* Knoxville, TN: Energy, Environment and Resources Center at the University of Tennessee.

Mathews, David. 1994a. *Politics For People: Finding a Responsible Public Voice.* Urbana and Chicago, Il: University of Illinois Press.

_____. 1994b. International Principles of Public Participation: Prospects and Challenges. Speech at Building Partnerships: Worldwide and at Home. Washington D.C.: Conference of the International Association of Public Participation Practitioners.

NUREG 1496. 1994. *Generic Environmental Impact Statement in Support of Rulemaking on Radiological Criteria for Decommissioning of NRC-Licensed Nuclear Facilities.* Washington D.C.: Government Printing Office.

Office of Technology Assessment. 1993. *Aging Nuclear Power Plants: Managing Plant Life and Decommissioning.* Washington D.C.: U.S. Congress.

Paulson, Jerry. Former Executive Director of McHenry County Defenders. Personal Communication.

Pollock, Cynthia. 1986. *Decommissioning: Nuclear Power's Missing Link.* World Watch Paper #69. New York, NY: World Watch Institute.

Thompson, Anthony J. and Michael L. Goo. 1994 "The Decontamination and Decommissioning Debate." *Radwaste.* (April): 32-41.

U.S. Congress. 1985. *U.S. Code, Congressional and Administrative News.* Washington D.C.: Government Printing Office.

U.S. Department of Energy. 1986. *The 1985 State-By State Assessment of Low-Level Radioactive Wastes.* Washington D.C.: Government Printing Office.

Yusko, James. 1993. "Radioactive Scrap Metals." *Proceedings of the Radioactive Scrap Metal Conference.* Energy, Environment and Resources Center at the University of Tennessee.

PART III

The Globalization of
Nuclear Power

Chapter 7

Nuclear Power and Postindustrial Politics in the West

Michael T. Hatch

Throughout much of the postwar period, economic prosperity in the West was fueled by exponential growth in oil consumption. As long as oil supplies remained abundant and cheap, this development occasioned little concern. But the 1973-74 energy crisis, with its skyrocketing prices and supply interruptions, sent shockwaves through the economies of the advanced industrial democracies in the form of inflation, recession, unemployment and balance of payments disequilibria. At the same time, such environmental problems as air and water pollution — largely caused by existing energy production and consumption patterns — were beginning to assume more prominent positions on the domestic agendas of many countries in the West. Nuclear power was widely seen as a major alternative to expensive, politically unreliable and polluting energy sources. The promise of clean, abundant and inexpensive nuclear energy, however, now appears to be ephemeral. Throughout the West, nuclear power is an industry in crisis, if not collapse. This chapter examines the factors that have brought nuclear power to this point.

The analysis focuses on the nuclear power programs of France, Germany, and the United States — arguably the three countries most influential in determining the past and future course of nuclear power in the West — and the political processes that shaped those programs. More precisely, it is argued that as the political divisions and values shaped by industrial society were challenged in emergent postindustrial societies, the corporatist political arrangements that characterized the nuclear programs of the U.S., Germany and France have become tenuous and difficult to maintain. When these arrangements gave way to a more pluralist policy process, political stalemate has threatened the viability of nuclear power (Hatch, 1991a).

Nuclear Power and the Political Process

The ideal typical pluralist model is characterized by a multiplicity of voluntary groups spanning a broad range of interests, all competing on a more-or-less equal basis for access to government decision makers and influence over public policy. Parliament and political parties provide relatively easy access points to the political process. Competition among interest groups, combined with open entry into the policy process for existing and potential groups alike, is said to lead to moderation in demands, compromise, and pragmatic negotiation (Braybrooke and Lindblom, 1963; Lindblom, 1968; Lindblom, 1959). The pluralist model is not without its detractors, however. According to the "overload" school, for example, pluralism has been the source of the "ungovernability" that has bedeviled major industrial democracies since the 1970s (Crozier, Huntington, and Watanubi, 1975). More precisely, democratic participation has broadened as previously passive or unorganized groups sought to establish claims on government. Rather than reflecting healthy competition among interest groups, however, this "overparticipation" overloaded the system both in terms of the number of demands as well as expectations (promoted in large part by competitive bidding among political parties). The result has been the incapacity of government to govern effectively (Berger, 1981; Lehmbruch, 1983; Manley, 1977; Lindblom, 1977; Schmitter, 1974).

Corporatism, in contrast, is viewed as an institutionalized, consensual pattern of policy making — one typified by close collaboration between the state and functionally organized interest groups in the formulation and implementation of public policy. When juxtaposed to pluralism, the interaction between interest organizations and government is distinguished by the central role of the state in shaping the content of policy. Moreover, certain interest groups are granted preferential access to a governmental policy process dominated by the bureaucracy, thereby enhancing the "governability" of modern, industrialized societies (Goldthorpe, 1984; Lehmbruch and Schmitter, 1982; Schmitter and Lehmbruch, 1979; Schmitter, 1981).

United States

The United States is generally viewed as the prototype of a pluralist political system, one characterized by a constitutional separation of powers investing Congress with considerable independence and power vis-à-vis the executive branch, a congressional committee system that fragments power even further, a federal structure that cedes considerable power to the states, and a politically benign climate for organized interests. In the first decades after World War II, however, nuclear power was shaped by a policy process more akin to corporatism than pluralism.

The state has been intimately involved in the development of nuclear power from its inception, given its roots in the wartime Manhattan Project. While the military initially dominated the program, civilian oversight was confirmed in the 1946 Atomic Energy Act with the creation of the Atomic Energy Commission

(AEC) and the Joint Committee on Atomic Energy (JCAE). Nuclear power remained a government monopoly over the next decade, but the 1954 Atomic Energy Act established private sector primacy in commercial applications for nuclear power. Nonetheless, the public sector continued to play a central role through the AEC — charged not only with the regulation but also the promotion of nuclear power — and the JCAE, the lone congressional committee responsible for all nuclear matters. The AEC and JCAE — along with the nuclear industry (which included electric utilities, such major reactor companies as General Electric and Westinghouse, and numerous engineering firms such as Bechtel) and the scientific community (largely employed or funded by the federal government) — came to form a powerful, relatively autonomous policy-making network that controlled nuclear policy for almost two decades.

By most accounts, an intimate, insular relationship developed between the nuclear industry, the AEC, and the JCAE — one that emphasized the promotion of nuclear power (Campbell, 1988: 78; Chubb, 1983: 92-93; Jacob, 1990: 27; Rosenbaum, 1991: 247). Nuclear policy was formulated largely within the AEC in close consultation with industry; any environmental or citizen interest groups that may have had concerns about nuclear power had little access to the process. With the JCAE holding a monopoly over nuclear affairs in Congress, the legislative avenue to the policy process was foreclosed to most outside interests. The ability to influence policy at the state or local levels was circumscribed by the doctrine of federal preemption, which meant that rules promulgated by the AEC on nuclear matters could not be overridden by state or local governments.

Out of these corporatist arrangements emerged a set of measures specifically designed to encourage a commitment to nuclear power by the private sector. Included in those measures were the Price-Anderson Act of 1957 — which limited industries' liability for a nuclear accident to $500 million (amended to $7 billion in 1988) — and large government subsidies, which took such forms as R&D funding,[1] production of nuclear fuel, and an agreement to accept utilities' spent fuel. By the mid-1960s, commercial orders began to show dramatic increases.

When the energy crisis struck in 1973-74, nuclear power appeared ready to assume a central position in America's energy future. In its official response to the crisis ("Project Independence"), the Nixon administration proposed that nuclear power provide 30-40 percent of U.S. electricity generation within 15-20 years (up from 4.5 percent in 1973), more than 50 percent by the year 2000. Coinciding with this greater public commitment to nuclear power, however, was an erosion in the institutional and procedural means to deliver on this commitment–the corporatist arrangements that had characterized nuclear power were increasingly under siege.

By the early 1970s, antinuclear activity had increased dramatically in conjunction with concerns about nuclear safety and criticisms of a policy-making process that allowed little voice for those concerns. Initially scattered and organized at the local level, opposition to nuclear power had become a national movement by the mid-1970s when such groups as the Sierra Club, Friends of the Earth, Natural Resources Defense Council, Union of Concerned Scientists, and Critical Mass became involved (Bupp and Derian, 1981: 132-135). With the antinuclear movement pressing its demands on all sides, the autonomy of the policy process began to disintegrate.

---

[1] Between 1946-59, the AEC provided $2.5 billion for reactor R&D and around $1 billion for the construction of facilities compared to $650 million invested by the industry. See Jacob, 1990: 25.

In 1974, the AEC was abolished. Its regulatory authority was given to the newly created Nuclear Regulatory Commission (NRC); the Energy Research and Development Administration (later changed to the Department of Energy or DOE) assumed responsibility for development. Though criticism of both institutions continued (the NRC for maintaining close ties to industry and the DOE for promoting nuclear power at the expense of more environmentally benign or less costly alternatives), the changes did have an impact. The NRC, for example, was now required to make its deliberations public, keep records of its proceedings, and allow the public to challenge agency decisions (Jacob, 1990: 49). Nonetheless, aside from greater transparency, the licensing process was still structured to favor industry (Chubb, 1983: 93-94). As a consequence, institutional reorganization within Congress took on greater significance.

The JCAE had demonstrated little sympathy for concerns raised by the anti-nuclear forces over environmental protection and safety. Many members of Congress also resented the far-reaching powers enjoyed by the JCAE. Following several resignations and electoral defeats within its ranks in the 1976 elections, the JCAE was abolished the following year. In contrast to earlier times, when the JCAE had buffered the AEC/NRC from many societal pressures, the NRC and DOE now had to deal with over two dozen House and Senate committees and subcommittees representing much broader constituencies.

Aside from greater access to the policy process through Congress, environmental groups found the courts a promising avenue to pursue their challenges to nuclear power. In the Calvert Cliffs decision, the Supreme Court ruled that under the 1969 National Environmental Policy Act (NEPA), an environmental impact statement was necessary before the AEC could license a commercial reactor. In other words, NEPA provided citizen groups (as well as states) an important instrument in their efforts to gain entry into the decision-making process.

Given the doctrine of federal preemption, state and local governments were clearly the less powerful actors when it came to nuclear matters. This is not to say, however, that they were powerless. States, through the rate-setting authority of their public utility commissions, could restrict the construction of nuclear power plants. They also had regulatory responsibilities for the non-nuclear environmental and siting aspects of nuclear projects. Moreover, the possibility of referenda at the state level allowed antinuclear activists to place such controversial issues as nuclear safety and radioactive waste on the ballot in several states.

Finally, as demonstrated by the decision by President Carter to defer indefinitely reprocessing because of proliferation concerns and his efforts to terminate the Clinch River fast breeder project, direct White House intervention had the effect of fragmenting further the policy process.

In sum, the corporatist arrangements that had characterized nuclear policy throughout much of the postwar period gave way to more pluralist arrangements by the mid-1970s. With the opening up of the policy process, opponents of nuclear power were able to slow its expansion substantially. Largely in response to concerns about reactor safety — and the ability of antinuclear forces to intervene in the licensing process — over 300 new AEC/NRC regulatory guidelines were issued between 1970-80. Licensing delays, design alterations, and retrofitting followed, pushing up the costs of nuclear power dramatically. When state public utility commissions were unwilling to allow those higher costs to be passed on to the rate payer prior to operation, the economic attractiveness of nuclear power declined further (Campbell, 1988: 85-87; Jacob, 1990: 52). By 1975, new orders for nuclear plants started to show a sharp decline; at the same time, cancellations of existing orders began to increase (twelve in 1978 alone).   In essence, an informal moratorium on the purchase of new reactors was in place by the late 1970s. The nuclear accident at Three Mile Island in 1979 only served to re-enforce a dynamic already in place.

Federal Republic of Germany

The Federal Republic — of all the large industrial democracies of the West — is said to most closely approximate a corporatist style of policy making (Hancock, 1989; Katzenstein, 1985). In contrast to the United States, the development of nuclear energy — of necessity — had an exclusively industrial (as opposed to military) orientation, given the sensitive political situation of West Germany in Europe after World War II. Nonetheless, the high cost of nuclear R&D, combined with long lead times for commercialization, meant that government would play an important role in nuclear affairs.   In May 1955, representatives from various electronic and chemical companies created a study group to look at prospects for nuclear research. That following October, the Ministry for Atomic Questions (after a succession of name changes now called the Federal Ministry of Research and Technology — BMFT or Bundesministerium für Forschung und Technologie) was established to give order to the relationship between private, state, and federal authorities in nuclear matters. Consultation between the Ministry and outside interests were formalized with the creation of the German Atomic Commission, composed of individuals from industry, the scientific community and government. Subsequently, a series of nuclear plans were formulated by the Atomic Commission — at first with only informal government involvement, later with official government participation. Through the coordinated actions of these various participants in the policy process, efforts to bring German nuclear power to the commercial stage culminated in the late 1960s. In April 1969, after strong encouragement from the federal government, the two major German companies involved in nuclear reactor development — Siemens and AEG — formed the Kraftwerk Union (KWU), a joint subsidiary created to supply the national market (Nau, 1974: 72-76, 85-87, 91-93). In other words, corporatist arrangements characterized the nuclear policy process prior to the energy crisis.

Since the mid-1970s, the overriding rationale of Germany's energy policy has been to reduce its dependence on imported oil (energy imported into the Federal Republic rose from six percent of total energy consumption in 1957 to 55 percent by 1972). Rapid expansion of nuclear power was seen as critical to the realization of this objective. Coalition governments from both the center-left (Social Democratic Party or SPD and Free Democratic Party or FDP) and the center-right (Christian Democratic Union or CDU and FDP) have held tenaciously to the conviction that nuclear power was essential if Germany were to achieve greater energy security; their ability to act on this conviction, however, waxed and waned.

In response to the actions of OPEC in 1973-74, the German government adopted an energy program calling for a reduction in West Germany's dependence on imported oil (from 55 percent of total energy consumption to 44 percent by 1985). The corollary of this goal was the rapid expansion of nuclear energy, from 1 percent of primary energy consumption in 1973 to 15 percent by 1985. This involved the construction of approximately 50 nuclear power plants.

Before the energy crisis, energy issues had not concerned most citizens in the Federal Republic. This ended in the wake of the events of 1973-74. Initial opposition to nuclear power took the form of local ad hoc citizens' initiatives (Bürgerinitiativen) organized to protest the construction of nuclear power in their region (Hatch, 1986: chapter 4; Nelkin and Pollak, 1981); however, with government plans for the construction of 45-50 nuclear power plants by 1985, concern over nuclear energy ceased to be a localized phenomenon. Public discussion was no longer restricted to such local concerns as thermal pollution or radioactivity escaping from plants, but rather quickly came to include questions about the long-term disposal of radioactive waste, the political, social, and economic ramifications of a "plutonium economy" in which spent fuel was reprocessed and fast breeder reactors were introduced, and the possible consequences for society of foregoing development of nuclear power.

Whereas the mass demonstrations of the antinuclear movement triggered the broadening public debate, a parallel and complementary course of action — less obtrusive but perhaps more effective in actually influencing the plans of industry and government — was redress through the legal system. As a consequence of cases brought before the administrative courts by nuclear opponents, work on three of thirteen reactors under construction had been halted by the end of 1977. Moreover, citizens' initiatives had begun to call for a moratorium on nuclear power plant construction.

The nuclear debate posed major problems for the political parties in power. Serious and deep splits were emerging within the SPD over nuclear power. Chancellor Schmidt and most SPD cabinet members were strong proponents of nuclear power; however, at the state and local levels, party leaders were feeling the effects of the antinuclear movement. The SPD's smaller coalition partner, the FDP, experienced the same sort of divisions as well (Hatch, 1986: Chapter 4).

Continual waffling of the political parties on the moratorium issue as a consequence of these divisions convinced many environmentalists that the established political parties were more a part of the environmental problem than of its solution. Beginning in early 1978, "Green" parties were established in various states to make nuclear power a major issue in the series of electoral campaigns for representation in state parliaments; by 1980, they had organized to contest national elections.

The Greens had a dramatic impact on the nuclear debate and the political process shaping that debate. Given the constitutional requirement that parties must receive at least five percent of the vote to be represented in parliament, they altered the outcome of several state elections, in some instances drawing off enough votes from the FDP to bring it below the five percent margin, at other times keeping the SPD and/or FDP from forming the government.[2] They set the tenor and content of certain campaigns, thereby compelling the established parties to address the issues that most concerned the environmentalists — this resulted at times in the regional SPD and FDP distancing themselves from the position of their own national government. Finally, they surmounted the five percent threshold at the state level in 1979, the national level in the 1983 federal elections.

Thus, the features unique to the German electoral system, in combination with the delicate, often precarious, representational balance among the established political parties, contrived to enhance the leverage of environmentalists in the political process and, at the same time, keep at center stage controversies surrounding government nuclear policy.

The major points of controversy centered on two important components of the nuclear program: reprocessing of nuclear waste and development of the fast breeder reactor. Of the two, the waste disposal problem had become the pivotal issue relating to the further expansion of nuclear power in the Federal Republic.

---

[2]   According to demographic studies, although Green supporters came from all parties, a larger proportion were drawn from the SPD and FDP than the CDU/CSU.

Central to the government's plan for nuclear waste disposal was the construction of a large, integrated facility designed to reprocess spent fuel rods and provide final burial for nuclear waste, all within a single site. The site selected by the federal government for further study was located at Gorleben in the state of Lower Saxony.

After over two years of delay, the CDU-led Lower Saxony government refused permission to proceed with the reprocessing plant. Though the CDU favored nuclear power in general and the Gorleben concept in particular, government leaders in Lower Saxony refused to support the federal government's program for waste disposal as long as that government's regional parties continued their opposition.

This decision created problems for the federal government, the greatest being its potential impact on the courts. With the integrated waste disposal concept apparently dead for the time being, officials feared that power plant construction could be delayed even longer; or worse, that the courts could demand a shutdown of plants already operating until the questions concerning waste disposal were satisfactorily resolved.

In sum, pluralist politics came to dominate the policy process through the late 1970s, resulting in political stalemate. Following the energy crisis, environmental groups rose to challenge the government's energy program — initially through public demonstrations and the courts, later through the electoral process. The political parties became critical participants in the debate over nuclear power as the prospect of defections to the Green parties threatened to keep them out of office or even out of parliament. Rather than narrowing the differences over nuclear power, however, divisions within the parties were exacerbated and tensions between state and federal governments surfaced. Under pressure from the grass roots, regional party organizations increasingly came into conflict with their national party leaders and state governments thwarted efforts of the central government to keep critical elements of the nuclear program on track.

France

As the effects of the oil embargo and four-fold price hikes of 1973-74 reverberated through the economies of the industrialized West, French officials proposed one of the most ambitious nuclear programs in the world, the overriding rationale being to reduce France's extensive dependence on increasingly expensive, insecure supplies of imported oil. The extent of the commitment to nuclear power was reflected in the figures contained in the program as it evolved through the 1970s:

- In the midst of the energy crisis in 1974, the government called for a speedup of the nuclear program, with construction to increase from 2000 megawatts (MW) in 1973 to 6000 MW through 1980. In 1975, this decision was formalized in a comprehensive program calling for nuclear energy to meet 25 percent of France's total energy needs by 1985 — up from only 2 percent in 1973. Accordingly, Electricité de France (EDF, the state-owned electric utility) was authorized in February 1975 to build 12,000 MW during 1976-77–approximately six plants per year.

- In 1976, it was decided to limit construction over the next two years (1977-78) to 5000 MW per year for reasons largely to do with slower growth in the demand for electricity. In the wake of the second oil crisis, however, the government announced a renewed acceleration of nuclear power construction — an added 5000 MW output every year — followed in 1980 by a revision of the government's long-term energy program.   By 1990, nuclear was to provide 30 percent of primary energy consumption, up from the 1978 level of 3.5 percent.

• Finally, and over the longer term, the fast breeder reactor (FBR) was to assume an ever more important role in nuclear power generation; from 1985 to the year 2000, two FBRs were expected to be ordered every three years.

In contrast to the United States and Germany, where efforts to implement similarly ambitious nuclear programs were continually frustrated, the French government was undeterred in its policy of rapid nuclear expansion. The reasons had to do with a corporatist policy process and the role of the state in that process.[3]

Since the advent of the Fifth Republic, the final decision on all major elements of France's energy program has been taken within an interministerial council headed by the President. But the major elements of the program were worked out beforehand within consultative committees, the most relevant for nuclear power being the influential PEON Commission (Commission Consultative pour la Production d'Electricité d'Origine Nucléaire). It was composed primarily of the highest-ranking officials in EDF, the CEA (Commissariat à l'Energie Atomique, a public body charged with the responsibility of developing nuclear energy), the nuclear industry (Framatome being the most important firm), the Plan, and the Ministries of Finance and Industry (Simonnot, 1978).

---

[3]   Given the historically prominent role of the state in French politics, the corporatist/pluralist distinction might foreclose a more promising approach to the analysis of politics in France — one that emphasizes the distinctive place of the state in the policy-making process. The position of the state in the bargaining process (whether as co-equal or as the more forceful of the participants), however, could be considered one of the dimensions of the corporatist model rather than a factor that invalidates it. Arguing the former, see Wilson, 1987: 241-70; the latter, see Hall, 1986: 270.

The state representatives in PEON were clearly the more powerful actors; and there was an overwhelming consensus within the state on the necessity of nuclear power for protecting France's national independence. But this did not imply that the nuclear industry was simply a rubber stamp. In the late 1960s, for example, French industrial interests helped influence the decision to abandon the French-designed gas-graphite reactor in favor of the American light water technology (Bupp and Derian, 1981: 60-69; Saumon and Puiseux, 1977: 146-149). It also did not imply the absence of conflict within the state over nuclear policy, the prime example being the dispute between the Ministry of Finance and EDF in the mid-1970s over the rate of nuclear construction. Nonetheless, given the state's ability to restrict participation in the policy process through its control over membership on the PEON Commission, dissenting voices, especially any urging caution about the rapid expansion of nuclear power, were conspicuously absent from the deliberations of PEON. Alternative points of access to the political process were foreclosed as well.

In contrast to the U.S. and Germany, where licensing procedures and the court system provided access to the policy process, France had a centralized licensing process dominated by state institutions (EDF and the Ministry of Industry). No application for construction permits was ever denied; and when decisions were appealed to the administrative courts, they met with little success.

Parliament, as well, was not a fruitful access point for antinuclear groups. Throughout the 1970s, it remained far removed from the conduct of nuclear policy.

The absence of parliament from the policy process, in part, reflected a general consensus among virtually all political parties on the major thrust of French energy policy. Both major parties on the right — the Gaullists (RPR) and the Republicans (PR) — were strong advocates of nuclear power; and only minor points differentiated the parties on the left from their opposition. More significantly, however, pressure on the political parties to respond to the concerns of nuclear opponents was reduced through the electoral system (a qualified exception being the Socialists or PS). In a single member district, two round voting system, candidates from newly created ecological parties had little chance of moving on to the second round, thereby minimizing the necessity of established parties to compete electorally for the environmental vote.

The quick implementation of the nuclear program was due in no small part to corporatist arrangements that limited participation in the policy process and the role of the state in those arrangements. This did not imply, however, that the state was completely disconnected from societal influences. The bureaucratic portion of the state, for example, must take its cues from the popularly elected political executive when dealing with areas of interest to the President. Moreover, as the experience of cohabitation (an arrangement where the president and prime minister are from different political parties) demonstrated, the powers of the Presidency are conditioned by the configuration of political parties in parliament. In other words, the ability of the state to play the role that it has in the political process rested on the acquiescence, if not the approval, of a majority party or a parliamentary coalition. By implication — given the powerful position of the state — any changes in the policy process most likely would require the conversion of a political party to the necessity of change, combined with the opportunity to put those changes in place. For many critical of the political process, and the policies on nuclear power produced by it, the Socialist Party became the receptacle of their hope for change.

In the face of pressures mounting from within as well as outside the party, the PS began to modify its position on nuclear power by the late 1970s. Newly formed ecologist parties did surprisingly well in the March 1977 municipal elections, receiving over 10 percent of the vote. For a PS contesting national parliamentary elections the following year, this development held potential dangers since studies indicated that ecologist voters were more likely to come from the left, specifically from the Socialists, than the right. In addition, the CFDT — the second largest trade union in France, the largest union in the nuclear sector, and the one most closely aligned with the PS — was beginning to voice certain reservations about the government's nuclear program because of its concern about the working conditions of its members.

Following the breakdown of its electoral alliance with the Communist Party in September 1977, the Socialists revised their nuclear position. While still affirming general support for nuclear power, the PS called for an eighteen to twenty-four month moratorium on nuclear plant orders and the immediate suspension of construction on the FBR Super-Phénix. Further modifications of this position followed in the wake of the Three Mile Island accident and in preparation for the 1981 presidential elections.

Limiting the expansion of nuclear power, no industrial use for the FBR, and an examination of possible alternative methods of waste disposal (i.e., whether or not to reprocess nuclear waste) were the major demands specific to the technology itself. Equally significant, however, were reforms proposed by the PS to open up the policy-making process. Juxtaposed to what they termed the "Bonapartist" tendencies of the Giscard government, the Socialists called for greater democracy and decentralization in the decision-making structures and procedures governing nuclear policy. Specific proposals included a call for a national referendum on the development of nuclear power and the creation of a Nuclear Law that would shift more power to regions, departments and municipalities as well as enable more control over the policy process by citizens and elected officials.

Once in office, the initial measures undertaken by the new Socialist government raised the hopes of nuclear opponents. The controversial Plogoff plant in Brittany was canceled outright, construction on five other reactors was frozen, and a parliamentary debate was initiated over the rate of nuclear construction, the future of the FBR, and the expansion of reprocessing facilities. In October 1981 the debate was concluded with a vote in the National Assembly, but little had changed as a consequence: six new LWRs were to be ordered over 1982-83, three fewer than planned by the Giscardian government because of lower energy requirement forecasts; the Super-Phénix was to be completed, with a decision on starting a series of FBRs for industrial use to be reserved for later determination; reprocessing capacity was to be expanded; and virtually nothing was done involving the Nuclear Law and a national referendum (Hatch, 1991b).

Nuclear Power Under
Siege After Chernobyl

United States

On 26 April 1986, the most serious accident in the history of nuclear power occurred when an explosion at the Chernobyl nuclear facility allowed radioactive materials to escape into the atmosphere. Though the political impact of the accident varied across states, repercussions soon followed as it rekindled intense political debate over nuclear power.

For nuclear power in the U.S., the Three Mile Island accident had a more immediate, direct impact on nuclear policy; Chernobyl — though perhaps inflaming such long-running controversies as Seabrook and Shoreham — simply re-confirmed the political stalemate that had been in place for over a decade.

In the area of licensing, two of the most celebrated cases were the Seabrook nuclear facility in New Hampshire and the Shoreham plant on Long Island, where the states of Massachusetts (which was within the ten mile radius of Seabrook) and New York refused to provide the evacuation plans mandated by Congress following Three Mile Island. In the absence of evacuation plans, both the Shoreham and Seabrook plants, though completed by the mid-1980s, were unable to begin operation. In an effort to overcome the impasse, a presidential order was announced in November 1988 that allowed the Federal Emergency Management Agency to devise and carry out emergency plans, thereby overriding local opposition.[4] For the Shoreham plant, however, the change was too late; it was closed before ever operating at full capacity. At Seabrook, the second unit had been canceled in 1986. In 1988, the utility applied for bankruptcy because the state utility commission denied a rate increase to recover construction costs before the reactor began commercial operation, but unit one did finally come on line in 1990.

A second major area of ongoing concern has been nuclear waste management. The absence of a working waste management facility became an important issue by the mid-1970s. In 1976, the California Energy Commission announced that it would not approve any more nuclear plants unless the utilities could specify fuel and waste disposal costs, an impossible task without decisions on reprocessing, spent fuel storage, and waste disposal. By the late 1970s, over thirty states had passed legislation regulating various activities associated with nuclear waste. Several states, for example, had passed statutes similar to California's. Others prohibited a waste management facility. Illinois had banned the storage of spent fuel from outside the state. And numerous states

---

[4]   The form of the regulatory change (a presidential order rather than legislation) combined with the timing of the announcement (delayed until after the elections), reflected the political sensitivity of the issue.

and localities had adopted regulations on waste transport (Jacob, 1990: 64). In 1983, the Supreme Court upheld California's right to decide if nuclear plants were justified on economic grounds, including the anticipated costs of waste management.

Legislative deliberations on the waste question began in 1980. The chief concern for many states was the possibility that they might be designated a potential site for a nuclear waste repository. Over the course of negotiations, public officials from several states were able to include provisions in the legislation that effectively eliminated them from consideration. Other states demanded that the bill stipulate the right of veto by any state chosen as a waste site; the veto could only be overridden by majority votes in both houses of Congress. Finally, a central element of the legislation was an agreement to build two facilities, the first in the West and the second in the eastern or central part of the U.S. (Jacob, 1990).

The Nuclear Waste Policy Act (NWPA) was passed in 1982. Among other things, it required the President to submit the choice of the first repository to Congress by 1987. Once NWPA became law, DOE designated nine potentially acceptable sites as candidates for further study. Almost immediately, several states passed legislation designed to impede the testing of prospective sites; and law suits were filed challenging the selection criteria (Campbell, 1988; Rosenbaum, 1987: 146).

In 1985, DOE nominated sites in Nevada, Washington and Texas for additional studies. Political pressures on the White House from politicians in each state followed. Environmental groups challenged the designations in the courts. When the group of potential sites for the second repository was announced in 1986, opposition from seven eastern and central states became so intense that DOE withdrew its proposal. By 1987, with the procedures prescribed by the NWPA breaking down, Congress set about to amend the legislation.

Leading the efforts to revise NWPA was Senator Bennett Johnston from Louisiana, democratic chair of both the U.S. Energy and Natural Resources Committee and the Appropriations Subcommittee on Energy and Water. With the support of the Reagan administration, he used the levers of power at his disposal to fashion a compromise that, in effect, left only one site available for testing.

The centerpiece of the strategy was to postpone any search for a nuclear waste repository in the East for twenty years, thereby sparing the seven states on the initial list. To get the support of states not on the list, Johnston reportedly threatened to cut off funding for states' water projects or suggested in letters to Senators that their own states might be selected if this proposal failed. A provision was included in the legislation stipulating that a waste repository could not be located under an aquifer. This eliminated Texas and Washington as potential sites (Novak and Kaplan, 1988: 22-23; *Wall Street Journal*, 1 December, 1987). The only site remaining was Yucca Mountain in Nevada — a state with one of the smallest populations and congressional delegations in the nation. The amended version of NWPA was adopted by Congress in December 1987.

Yucca Mountain was originally scheduled to open in 1998. In 1989, DOE announced that the repository would open, at the earliest, in 2010. In 1993, the General Accounting Office estimated that the site investigation alone could take at least five to thirteen years longer than planned.

It is clear that politics rather than science drove the policy process that determined the choice of the site. What is less clear is how the site evaluation will proceed (Greenberg, 1993). Since no other sites are currently under active consideration, the underlying assumption appears to be that Yucca Mountain will be suitable for the permanent storage of high-level radioactive waste — an assumption that may not be on solid footing.[5] Should subsequent studies find the site unsuitable, the government is left with few alternatives in its efforts to resolve the uncertainties about nuclear waste management. This, in turn, could impinge on issues of growing salience to the nuclear industry — licensing renewal and decommissioning.

When he became chairman of the NRC in 1991, Ivan Selin announced that licensing renewal would be his top priority. The licenses of 49 of the 109 reactors in operation in the U.S. will expire over the next twenty years. Originally granted for forty years, the NRC has suggested that operating licenses, under certain circumstances, could be extended for another twenty years. According to the nuclear industry, however, DOE's ability to accept spent fuel will have an important impact on the license renewal question. That is, utilities are confronted with the question of how on-site spent-fuel storage will be paid for — and by whom — given that DOE will not be able take spent fuel in 1998 as originally planned (Bretz, 1994). Moreover, the costs of repairing and maintaining plants past their original life span may be very high. However, in light of recent decisions to close several reactors before their licenses had expired,[6] the more immediate question facing many utilities is one of premature shutdowns.

---

[5] Several potential problems have been raised that could eliminate the use of Yucca Mountain, including volcanic activity, close proximity to the nuclear bomb testing site in Nevada, and water flowing through the rock formation. See *New York Times*, 17 January 1989.

[6] The five reactors are Yankee Rowe, San Onofre, Rancho Seco, Trojan, and Fort St. Vrain.

Finally, the issues of licensing renewal and premature shutdown increasingly have become entangled in questions surrounding decommissioning.   Estimates for the cost decommissioning at this early stage of the process appear to vary considerably.   In 1988, the NRC put the figure for decommissioning large plants at $105 million to $135 million; an industry publication from 1992 estimated the costs would range from $150 million to $225 million, depending on the method (prompt dismantling or entombment) (Bretz, 1992: 48; Greenberg, 1993: 85).   The projected costs for decommissioning the Yankee Rowe (at 135 MW, a relatively small reactor) and Fort St. Vrain (330 MW) plants, on the other hand, are $247 million and $330 million respectively (Greenberg, 1993: 85).   In other words, when dealing with procedures for which there are few precedents to draw on, cost estimates are very unreliable.   The social and political ramifications of decommissioning, however, are no less daunting.

The volume of radioactive waste from the decommissioning of the reactors currently operating in the US is estimated at 81.5 million cubic feet (*Nuclear Waste News*, 21 July, 1988).   If the method of decommissioning is dismantlement, this will require permanent disposal sites and decisions regarding transportation routes.   It is unlikely that citizens would passively accept the transport of such large amounts of nuclear waste through their local communities and states.   The option of entombment, on the other hand, may be no less welcome to citizens facing the permanent presence of decommissioned reactors in their communities.   Given the access points provided various interests through the pluralistic policy process, the political uncertainties surrounding nuclear power are unlikely to disappear anytime soon.

Federal Republic of Germany

Following the refusal of the Lower Saxony government to proceed with the reprocessing plant in 1979, a joint federal/state resolution was formulated to provided a set of measures designed to satisfy the requirements for dealing with nuclear waste until a permanent solution was decided upon. With this formula, the federal government felt that the final impediment to further licensing of nuclear power plant construction had been removed.

Further steps were initiated in the early 1980s to "streamline" nuclear plant licensing procedures with the adoption of such measures as standardization of nuclear plant designs, unification and simplification of the licensing application procedures, and a reduction in the number of stages required in the approval process. Though criticized for the way these measures were adopted (only interests favoring nuclear power were represented, parties and parliament were circumvented) and the limitations placed on citizens' ability to appeal decisions made during the licensing process, the measures had the desired effect. In February 1982 — for the first time in four years and after a delay of five years — construction on three new nuclear power plants was authorized by the Interior Ministry in "convoy." That is, with the three plants following a standard design, licensing was to proceed more or less simultaneously in all states, with clearance in one state making clearance in the others automatic.

In fall 1982, the FDP switched coalition partners to form a government with the CDU/CSU (the CSU, or Christian Socialist Union being the sister party of the CDU in Bavaria). Since the CDU/CSU was not as threatened by electoral defections to the Greens, Chancellor Kohl was less encumbered by the internal divisions that, at times, had bedeviled Chancellor Schmidt's efforts to move forcefully on nuclear power questions. Accordingly, his government proceeded to address two additional issues that had been constant sources of friction within the parties and the center-left coalition as well as between federal and state governments: the decision was made to complete construction on the controversial fast breeder reactor at Kalkar in the state of North Rhine-Westphalia; and preliminary approval was given for the construction of a reprocessing plant, to be located at a site in Bavaria (Wackersdorf). In other words, after years of paralysis and equivocation, the major impediments to further expansion of nuclear power appeared to have been removed. And then came Chernobyl.

Despite government efforts to reaffirm the unequalled safety of German nuclear facilities, the public was not reassured, as evidenced in opinion polls that showed a consistently high level of opposition to nuclear power.[7]   Further undermining public confidence in the claims by government and industry about the rigorous safety standards applied to German nuclear facilities were a series of scandals in the late 1980s: reports about illegal shipments of fissionable materials to Pakistan and Libya; irregularities in the handling of nuclear waste and millions of marks in bribes to cover up the irregularities; and reports that German firms made illegal shipments of equipment for nuclear power plants to Pakistan, India, and South Africa.

Chernobyl also revived the antinuclear movement as large and emotional protests were staged throughout Germany that, at times, took on a civil war-like atmosphere reminiscent of the violent clashes between police and demonstrators from a decade earlier. When the frequency and intensity of the demonstrations fell off after several months, anti-nuclear forces pressed their concerns through the licensing process. At the hearings required for the licensing of the proposed reprocessing plant at

---

[7]   Approximately 70 percent of the population rejected nuclear power (*Der Stern*, 28 April 1988). More significantly, the level of opposition increased rather than abated in the years after Chernobyl. When questioned whether they rejected the construction of new nuclear power plants *and* whether they wanted the existing nuclear plants to be shutdown over a "transition period," 47 percent of the population answered affirmatively in November 1986, 57% in April 1987, and 63 percent in February 1988 (Wirtschaftswoche, 22 April 1988).

Wackersdorf, for example, over 800,000 objections to the facility were submitted.

Parallel to the growing public disaffection with nuclear power was a programmatic debate taking shape among political parties and interest groups over the call for an "Ausstieg" or exit from nuclear power. At the forefront of this debate were the Greens, who demanded an immediate shutdown of all nuclear facilities. This position found considerable resonance among the voters, as evidenced by their electoral performance in the January 1987 federal elections where they received 8.3 percent of the vote (up from 5.6 percent in 1983). The Social Democratic Party also moved quickly to contribute its voice to the debate over Ausstieg. The party set up a commission (the Hauff Commission, named after its chair) to study the question of future German energy supply without nuclear power. Following the recommendation of the commission, the SPD adopted the position that all forms of nuclear power should be eliminated over a ten year phase-out period.

In the aftermath of Chernobyl, Chancellor Kohl created a new Ministry for the Environment and Reactor Safety, in part, to reassure citizens concerned about the safety of nuclear facilities in Germany as well as emphasize the CDU's commitment to the environment. At the same time, the CDU leadership reaffirmed their unwavering support for nuclear power. Despite such affirmations, however, the debate over nuclear power and Ausstieg gradually began to intrude into the ranks of the CDU. By 1988, patterns similar to those found in the SPD when it headed the government started to appear. Among the reasons were that party members and voters also shared the doubts and concerns about nuclear power raised by the Chernobyl accident. The sum effect was a leadership caught between its commitment to nuclear power and the internal pressures from party members acutely aware of the impact the nuclear issue can have on electoral outcomes. When combined with the calls for Ausstieg by the SPD and Greens, questions about the political viability of nuclear power resurfaced with a vengeance.

One of the first areas of nuclear policy to be affected by the changed political climate after Chernobyl was the government's official program on reprocessing. Ironically, the electrical industry itself spiked that plan when it canceled the reprocessing project at Wackersdorf in April 1989. Giving as reasons the rising estimated costs for the facility and the uncertainties surrounding the licensing process, the German companies concluded an agreement with the French company COGEMA to reprocess their spent fuel. In the wake of the decision to terminate Wackersdorf came questions about the continuing validity of the government's waste disposal concept.

The most immediate political effects of this decision were felt in Lower Saxony, since the termination of Wackersdorf again left Gorleben as the sole site under active consideration for long-term nuclear waste management. Approval by the CDU-led state government in early 1990 to begin construction on a nuclear waste conditioning facility at Gorleben added a further element to the controversy as elections approached later that year. Throughout the campaign, the SPD and Greens raised the issue of nuclear power and waste disposal, much to the discomfort of the governing CDU and FDP; and with their election victory in May 1990, the new Red-Green coalition made stopping Gorleben a top priority in the government's announced policy of Ausstieg.

This election had been preceded by an SPD victory in the May 1988 elections in Schleswig-Holstein, where the party had run on a policy of Ausstieg, promising that if it received an absolute majority, operating licenses would be withdrawn from the three nuclear reactors located in the state. They were subsequently joined by the SPD-led government in the neighboring state of Hamburg in fulfilling their campaign promise. The strategy of the governments was to bring about an Ausstieg through the strict application of the laws governing nuclear power; however, efforts in all three states to implement the strategy have been limited by federal jurisdiction in nuclear issues.

One major area where the federal government has capitulated to state resistance in nuclear matters was the long unresolved issue of the fast breeder reactor. At its inception, the FBR represented one of the few possibilities for the Federal Republic to drastically reduce dependence on foreign energy sources over the long term. The plutonium produced in the operation of light water reactors — once extracted by reprocessing — was to provide the fuel for the FBR which, in turn, would produce more plutonium than consumed. By the late 1970s, however, concerns about the cost of the program and the safety of the technology, combined with growing doubts about the creation of a "plutonium economy," led to increasing tensions between the SPD-led federal government and the SPD government of North Rhine-Westphalia (NRW), site of the FBR prototype under construction at Kalkar.

Though construction delays resulted from the less than enthusiastic support of the NRW government, construction moved ahead. By 1986, Kalkar was approaching completion and increasing pressure was being applied by federal authorities (now controlled by the CDU) for the NRW government to license the loading and startup of Kalkar. Nonetheless, the state government stood firm in its refusal to issue any further licenses. Among the reasons given were the numerous (over 2,000) technical changes undertaken by the builders without approval from the licensing agency and the necessity of re-examining the safety of the FBR technology at Kalkar in light of the Chernobyl accident. Supporters of the FBR suspected political motives over legal considerations as the primary reasons for the government's opposition to the Kalkar reactor, but proponents too were hard pressed to justify the operation of a reactor already considered to be technically obsolete and an economic white elephant (approximately DM 7.5 billion had been spent at Kalkar). In March 1991, the federal government announced its decision to pull the plug on Kalkar, citing obstruction by the NRW government as the reason. This was not the sole factor, however. Federal officials could have overridden the state and brought the Kalkar facility into operation. With a good share of the German public rejecting nuclear power by this time, the federal government clearly had little appetite for further battle in this arena, its pro-nuclear protestations notwithstanding.

By the early 1990s, the cumulative effect of these developments left proponents of nuclear power, especially those in the private sector, increasingly skeptical about the viability of the industry unless some kind of an "energy policy consensus" could be achieved — that is, a broad-based political understanding on nuclear power that went beyond party lines and was accepted by the relevant social groups. The first soundings on possible areas of consensus began in fall 1992 in private discussions between the directors of two large electric utilities (VEBA and RWE) and Gerhard Schröder (SPD), Minister President of Lower Saxony.

At the heart of the discussions was a proposal by VEBA and RWE to replace existing nuclear reactors with coal and natural gas plants at the end of their operating life. The length of that operating life was to be specified in the agreement; industry thinking in this regard was to extend plant operation from the original 25 years to 40 years except for the older, more obsolete reactors. In other words, nuclear power was to be phased out, though over a longer period than specified by the SPD in their Ausstieg policy (the last reactor would be scheduled to close in 2029).

An additional issue for industry was to keep the nuclear option open over the longer term for the newer generation of reactors that was to incorporate features designed to make them much safer. As an inducement for the SPD and the Greens, RWE and VEBA offered to end reprocessing, conditioned on a change in the law making direct, final storage the only form of waste disposal. Under the direction of the Ministry for the Environment (BMU), consensus talks began in March 1993, using as the starting point the proposal developed by VEBA, RWE, and Schröder.

The negotiating group was composed of representatives from the federal and state governments along with members of the political parties from those governments; an advisory working group representing the electric utility industry, trade unions, and environmental groups was also established. As talks progressed, areas of potential agreement did emerge, but they also highlighted several points of contention within as well as between the various groups:

- Despite initial friction between the CDU-led BMU and FDP-led Economics Ministry over which was to lead the negotiations, the federal government took a position supporting the replacement of nuclear reactors at the end of their operating life, the construction of new nuclear reactors in the future if they did not endanger the environment, and final storage of spent fuel rods without having to be reprocessed.

- Industry's primary objective was to secure its investment in nuclear power by allowing present nuclear power plants to operate free from further challenges; however, utilities were divided over the possibility of Ausstieg. Utilities in the south such as Bayernwerk relied more heavily on nuclear power than the northern utilities of VEBA or RWE. In addition, they were far from the coast, thereby limiting their ability to import cheap coal. In other words, a return to fossil fuels would be more difficult for them than their northern counterparts in terms of inexpensive alternatives. The Bavarian state government and the CSU, its governing party, aligned with the utility in opposition to an Ausstieg, arguing the necessity of nuclear power for economic but also environmental reasons in light of concerns about global warming. KWU/Siemens, the only German firm building nuclear reactors, also rejected Ausstieg as well as an end to reprocessing, since it had invested approximately DM 1 billion to construct a plant that processed plutonium into mixed oxide (MOX) fuel elements.

●   For the SPD, there were several attractive features to the proposal–Ausstieg, exit from a plutonium with an end to reprocessing and the production of MOX; however, other issues divided the party. For a majority of the party leadership, leaving open the option of future nuclear plant construction was unacceptable; the group around Schröder, among them trade unionists, argued that later generations should be allowed to decide about the acceptability of new reactor types. Some were suspicious of Schröder's motives since, in conjunction with an agreement on the direct storage of nuclear waste, alternatives to the Gorleben final storage site were to be studied, thereby ridding him of the unending Gorleben problem. The SPD-led government in NRW was concerned that an interim storage site located there (Ahaus) might become permanent by default; the CDU-SPD coalition government in Baden-Würtemmberg was concerned that granite formations located in its Black Forest might become the alternative to Gorleben. Further, the major coal-producing states of NRW and Saarland, both led by SPD governments, were concerned that gas and imported coal–rather than expensive domestic coal–would replace nuclear power; the federal government, on the other hand, threatened to make coal subsidies contingent on SPD support for construction of the new nuclear reactor prototype. Finally, important elements within the party argued that a return to an energy policy consensus must emphasize energy conservation and increased energy efficiency.

● Within the Greens, party leaders in the governing coalitions of Lower Saxony and Hesse (the most prominent being Joschka Fischer, the Environment Minister in Hesse) generally favored participation in the discussions. While opposed to new nuclear plant construction, the timing for Ausstieg was negotiable, especially when they might be able to rid themselves of the Gorleben problem and the troublesome MOX plant, the object of ongoing legal battles between local anti-nuclear forces, Siemens, Hesse and the federal government. Others in the party refused to entertain even the possibility of compromise in the matter of nuclear power, urging immediate withdrawal from the talks.

By early July, the Greens had left the talks. The reason given was the federal government's unwillingness to enter into a real dialogue with opponents of nuclear power. In late October, the talks collapsed; the decisive factor appeared to be the SPD's rejection of the future use of nuclear power. Following the October 1994 federal elections, the CDU/CSU and FDP reaffirmed in their new coalition agreement support for the future construction of nuclear power plants and called for a resumption of negotiations on an energy consensus. Schröder too is reportedly attempting to again mediate an agreement between nuclear opponents and supporters (*Der Spiegel*, 28 November 1994). Given the deep divisions that continue to run through German society, whether such talks resume, let alone result in success, remains to be seen.

France

While many nuclear opponents and members of the ecological movement had placed their hopes for change in a PS victory, disillusionment and resignation set in after the 1981 elections as it became apparent that nuclear policy under a Socialist government differed little from its predecessors. The 1986 nuclear accident at Chernobyl, however, re-invigorated antinuclear activists.

In the immediate days after Chernobyl, the official position was that the radioactive cloud sweeping across parts of Europe missed French territory completely. A subsequent revision followed, admitting that the cloud had passed over France, but since it hadn't rained, no radioactivity fell on the land. It later became known that the radioactive cloud had been carried over practically all of France — not once, but twice — depositing doses of radiation in many areas that measured up to 400 times the normal level. More damaging, however, was the revelation that this was known to the CEA and that it chose not to make the information public. The manner in which the information was handled became the lightening rod for many of the concerns raised by Chernobyl.

Criticism focused on a system in which information about nuclear safety was too concentrated and centralized in the hands of the state agencies attached to the CEA, whose organizational task was to promote nuclear power. A proposal to redress this problem was put forward by the "Office parlementaire d'évaluation des choixes scientifiques et technologiques" (the Rapport Rausch). It called for the creation of an independent national agency for nuclear security and safety. The report was subsequently buried in the face of strong opposition from the CEA and a government cohabited by parties with even fewer doubts about nuclear power than the Socialists. This, however, did not end the revitalized debate over the nuclear issue.

While French energy experts and government officials had issued reassurances after Chernobyl that such an accident could not occur in French reactors, distrust of officials running the nuclear program grew among the French populace as state obfuscation and lies about the effects of Chernobyl in France became known. Subsequent "mishaps" at French nuclear facilities served to further undermine public confidence: workers at the Bugey plant, for example, were exposed to radioactivity; uranium hexafloride escaped from the Tricastin enrichment plant, injuring several technicians attempting to stop the leak; and at the Super-Phénix plant, more than 20 tons of liquid sodium (which explodes on contact with air) escaped from the cooling system into a secondary vessel before the leak could be located.

In addition to the concerns about nuclear safety provoked by these incidents, other problems confronting nuclear officials kept the nuclear issue in the public eye. For one, the strategy of constructing reactors in standardized series led to several "teething problems" in the latest generation 1,300 MW reactors, resulting in numerous shutdowns for repairs; the cost of repairs and lost power production approached FF 1 billion. A related problem has been an "epidemic" of generic cracks discovered in the vessel heads and generator tubes from an earlier generation of pressurized water reactors; costs for maintenance are estimated to be around $80 million per reactor. A drought in summer 1989 eventually led to cutbacks in electricity production because of a lack of water for the reactors' cooling systems. On the other hand, for several years prior to 1989, numerous reactors had to be shut down because of overproduction, with estimates of surplus generating capacity ranging from 5 to 12 reactors. The expense of idle capacity was further compounded when financing for the construction of these surplus reactors was taken into account (construction costs per reactor is approximately FF 1.7 billion; EDF's long-term debt in 1988 was around FF 230 billion). Finally, aside from safety concerns, Super-Phénix has been a financial drain on public resources when shut down for repairs as well as when operating (the cost of electricity produced by the FBR is estimated at three times the market price) (*Die Zeit,* 28 April 1989; *Frankfurter Rundschau,* 18 April 1988; *Suddeutsche Zeitung,* 20 January 1988; Greenpeace news release, 7 April 1994).

One effect of these assorted problems has been a growing mistrust of those responsible for the nuclear program combined with an increasing disaffection with nuclear power that now encompasses a majority of the French population.[8] Another related development was the revived support for the ecologist/green parties, as demonstrated by the surprisingly strong showing of the ecologists in the 1989 European elections (10.5 percent) and the regional elections in 1992 (14.7 percent). Whether these developments are of more general significance for French nuclear power and the policy process is open to interpretation.

The electoral support for the Greens could be viewed simply as a method of registering broad discontent with current conditions since the outcomes of European and regional elections are taken less seriously than those in national elections. But even if these electoral outcomes were a true reflection of public conviction, the ability of the Greens and their sympathizers to translate this sentiment into a change in policy is undermined by the electoral system. For example, though an electoral pact including the Greens and Génération Ecologie received 10.5 percent of the vote in the 1993 national parliamentary elections (not including approximately 3 percent cast for other "green" candidates outside the alliance), neither party captured a single seat. On the other hand, if growing concerns about environmental issues, including nuclear power, are the more permanent features of a "postindustrial" landscape in France *and* if green parties are able to

---

[8]   As reported in *Le Monde*, 27 October 1986, 52 percent of the population were hostile to the construction of new nuclear reactors and 64 percent thought those in positions of responsibility didn't tell the truth. A poll reported in *L'Express*, 11 June 1987, indicated that opposition to continued construction had increased to 58 percent; when asked if the dangers of nuclear power were unacceptable, 49 percent said yes; when asked if a Chernobyl were possible in France, 76 percent said yes. According to an opinion poll conducted in 1994, 63.9 percent of the French population opposed any further growth in nuclear power (*Inter Press Service,* 30 March 1994).

control a sizable portion of the vote, they may be able to have a considerable impact on electoral outcomes. Elements within the PS certainly seem to have taken this threat seriously in recent years.

In the early 1990s, with the blessing of President Mitterrand, Minister for the Environment Brice Lalonde established a rival green movement, Génération Ecologie. The intention, in part, was to attract Socialists who had gone over to the Greens. As the 1993 national elections approached, the PS had become increasingly concerned about Génération Ecologie, not just the Greens, poaching its voters. Facing an electoral debacle, the Socialists tried to woo the ecologists during the run-up to the elections in the hope that they would back PS candidates in the second round. While a wholesale shift to the PS in the second round may have been able mitigate the magnitude of the Socialist loss, no such shift occurred, thereby assuring a crushing defeat for the Socialists.

How is one to evaluate the status of nuclear power in light of these recent events? To a large extent, the current program continues to reflect certain compromises among the various interests represented in the corporatist policy process:

- Nuclear reactors continue to be built at a rate greater than necessary to meet energy needs for reasons that had to do with the demands of the nuclear construction industry. Due largely to concerns about burgeoning debt and huge surplus generating capacity, however, EDF announced in 1994 that no new orders would be placed before the year 2000.

- In the years since its completion in 1986, Super-Phénix — championed by the CEA — has operated only intermittently, a total of six months in the first six years. Concerns about repeated sodium leaks led to its shutdown in July 1991. In early 1994, the government approved the FBRs restart, but with a change of mission. Rather than operate as a prototype for electricity generation, its main purpose was to research the burning of plutonium, thereby reducing long-life radioactive waste produced by reprocessing.

●   Construction on two large nuclear installations has continued, a MOX fabrication plant at Marcoule and a spent fuel reprocessing plant at La Hague. Both are scheduled to be brought into industrial service during 1994, though the need for them may be diminished in light of the decision regarding fast breeder technology.

There are, nonetheless, recent incidents that at least raise questions about the possibility of change in the policy process.

In 1987, the state selected four sites to be considered as the final repository for nuclear waste. Initially this decision was met with the passive resistance of the local population which felt excluded from a policy-making process dominated by state agencies located in Paris. In December 1989, however, the nature of the conflict changed qualitatively. Protests escalated, demonstrations grew larger and environmentalists were joined in their opposition by local politicians from the major political parties, all making common cause against the siting of a nuclear waste facility in their locality. On February 7, 1990, Prime Minister Rocard announced that all work at the four sites would be halted for one year. He then solicited two studies, one from the "Office parlementaire d'évaluation des choixes scientifiques et technologiques" and a second from the independent "College de la prevention des risques technologiques."

The central point of the report from the parliamentary body was that parliament rather than the "nucleocrats" should have the last word where issues relating to nuclear waste were concerned;

moreover, local representatives should have a voice in the decision. A similar message was delivered in the second report: the social dimension of nuclear waste disposal had been underestimated. Inspired to a large extent by the proposals put forward by the parliamentary office, a law was passed in 1991 requiring parliamentary approval for waste management policy. It also established a "waste mediator" to negotiate with local communities over the right to undertake research and a 15-year cooling off period before a final site could be chosen. In 1993, the mediator — Christian Bataille, who had been selected from parliament — submitted a report to the Ministers of Industry and the Environment. The following year, the government selected four different geological areas for further study.

The decision to stop work on the nuclear waste projects — along with the subsequent handling of the waste question — may be an acknowledgment that a closed political process is increasingly dysfunctional in certain policy spheres. In other words, it is becoming more difficult to impose state policy on a populace that has been allowed little participation in the process and the political costs of doing so could be quite serious.

Conclusion

In 1994, 109 nuclear reactors were operating in the United States — by far the largest number of any country in the world — supplying approximately 20 percent of the nation's electricity. Clearly, the U.S. nuclear industry is not dead. It is, however, in crisis. Over 130 reactors have been canceled since the mid-1970s; and there has not been a single new order since 1978. From the late 1970s, public opinion has consistently opposed further expansion of nuclear power. As argued earlier, these developments have reflected, and been reflected in, the transition from the corporatist arrangements that had insulated policy makers from societal pressures in the first decades after WWII to a more pluralistic policy process that has resulted in a stalemate over nuclear power. Whether this stalemate will be broken any time soon is subject to debate.

In recent years, many proponents of nuclear power have looked to two developments that will enhance nuclear power's public acceptance: growing concerns about global warming and next generation nuclear reactors that incorporate a new passive safety design.   Nuclear power, it is argued, makes good environmental sense when confronted with the ominous threats posed by global warming, since no fossil fuels — the major contributor of greenhouse gases — are burned.   Further, opposition to nuclear power based on concerns about reactor accidents will be less tenable with the new passive safety designs.

Combined with the anticipated growth in public support for nuclear power is a strategy to encourage utilities to again commit to a nuclear future. Part of this strategy is to get pre-order design certification for standard reactor designs; NRC approval for the passively safe reactor designs are expected by 1995. A second element was the adoption in 1989 of new siting regulations which allows utilities to receive site approval before actual construction begins and allow the approved sites to be held ready for up to twenty years ("site-banking"). Finally, the National Energy Policy Act of 1992 approved one-step licensing procedures for new nuclear reactors which, in essence, means that all safety issues are to be litigated before construction begins, thereby reducing the possibility of costly delays once construction is underway. All told, this strategy represents an effort to carve out policy space buffered from the vagaries of the pluralistic policy process. Whether these efforts are sufficient to renew utility interest in nuclear power remains an open question.

Germany presently has the fourth largest nuclear generating capacity in the world, behind only the U.S., France, and Japan. The 21 reactors in operation provide 34 percent of the country's electricity supply; however, there have been no new orders since the late 1970s. In other words, the future for nuclear power in Germany appears far from assured, despite the continued support of successive governments from both the center-left and center-right. As in the U.S., German proponents of nuclear power have hoped that the growing political salience of global warming would provide the impetus to redirect nuclear policy. And, in fact, Germany has announced its intention to reduce carbon dioxide emissions 25-30 percent by the year 2005, perhaps the most demanding target of any advanced industrial democracy. Though opinions divided on the role of nuclear power, a broad consensus did emerge from the debate on how to achieve this target: drastic energy conservation and greater energy efficiency — not more nuclear power — was the key issue in fighting global warming (Hatch, 1993).

All told, the combination of forces presently arrayed against nuclear power — and the access those forces have to the policy process — does not bode well for the political viability of the technology. The reasons for this have much to do with the divided and diffuse nature of public power in Germany. Most obvious in its effects on nuclear policy is a type of federalism in which substantial power and authority reside at the state level. Further, the courts — operating primarily at the state and local levels — when combined with provisions for judicial review, have allowed nuclear opponents to exercise considerable influence in the policy process. Finally, in the party and electoral systems, where a single party rarely gains an absolute majority in national elections, the small parties are potential coalition makers (and breakers), and the five percent requirement remains a constant in electoral calculations — a relatively small number of voters for whom nuclear issues are very important can determine electoral outcomes, thereby giving them influence well beyond their numbers.

In many respects, the French nuclear program represents a remarkable achievement. Dependence on imported energy has dropped from levels approaching 80 percent in the mid-1970s to around 50 percent today, in large part because nuclear power has expanded to cover approximately 30 percent of primary energy consumption. Fossil fuels are of marginal importance in the electrical sector, with nuclear reactors now generating almost 80 percent of all electricity in France. Ironically, precisely those factors that help explain these exceptional accomplishments may now be leaving the program more vulnerable than ever before.

The corporatist arrangements — in which the French state both took the lead in, and served as gatekeeper to, the policy process — greatly facilitated the rapid expansion of nuclear power. State officials cavalierly dismissed the importance or usefulness of public participation in the political process, claiming technocratic competence and demanding public trust. Those holding elected office provided few antidotes to this perspective. The deceptions of Chernobyl and the various malfunctions since, however, have had their effect — a more cynical populace, increasingly skeptical about the claims of state officials in nuclear matters.

So long as problems in the nuclear sector remain relatively minor, this skepticism should not present any great difficulties. A major accident at a French nuclear facility is another matter. The problem of public acceptance of nuclear power would become more than simply a nuisance. Moreover, an accident in one of the reactors would create tremendous public pressure to shut down at least all the reactors of the same type since the cause may be due to a flaw in the series design itself. Given the role nuclear power now plays in France's overall energy supply, such a decision would have far-reaching financial, economic as well as political repercussions.

There are indications that these potential dangers are not lost on officials in positions of influence. One example was a recent internal report of EDF that, for the first time, did not exclude the possibility of a major accident (*Le Canard Enchaine,*

14 February 1990). Another was the creation of an expert study group headed by Philippe Rouvillois which submitted its report to the government in 1989.[9] Among the more important points were two:

- Anticipating the events of December 1989/January 1990, the study emphasized the importance of public acceptance in the question of final storage for nuclear waste; it also concluded that those most responsible for French nuclear power had not adequately informed the French public in this area.

- While implicitly critical of the dominant position of the nuclear lobby–CEA, EDF, Cogema, Framatome–in the policy process, the report placed primary responsibility on the government for allowing nuclear interests too much leeway in making policy. That is, rather than give clear political direction, the government limited itself simply to the implementation of earlier decisions.

In light of the problems related to declining public acceptance of nuclear power and the growing salience of environmental issues within France, the new center-right government initiated an unprecedented "national debate on energy and environment" in 1994. Between May and July, virtually all regions of France held regional debates. During the months of September and October, six roundtable discussions — structured to permit an exchange of opposing views — were organized around major themes emerging from the regional debates. A final debate was to be conducted by the French parliament in December. Though this debate was not expected to precipitate a fundamental change in the French nuclear program (*Le Figaro*, 9 May 1994), it

---

[9]   In July 1989, Rouvillois was appointed director of the CEA. Though suppressed, the "Rapport Rouvillois" became public when leaked to the newspaper *Liberation* in March 1990.

does perhaps demonstrate a growing awareness of the need for greater transparency in the policy process and for more control over nuclear policy by elected officials.

To conclude, pluralist politics certainly contributed to the stalemate over nuclear power in the United States and Germany over the past two decades. But, would a corporatist policy process that systematically excluded significant segments of the population from effective participation have represented a better alternative? In conjunction with the corrosive effects of such arrangements on the legitimacy of government actions in postindustrial societies, one might speculate about the viability of the nuclear industry today in the absence of pressures placed on it through the pluralist policy process for safer operation. In the case of France, what has become increasingly apparent is the declining efficacy of corporatist arrangements dominated by the state at the exclusion of important sections of the population. Ironically, the future viability of nuclear power in France may hinge on the ability to make permanent the more open political processes suggested in the recent national energy debate and the procedures developed to deal with nuclear waste.

References

Berger, Suzanne (ed). 1981. *Organizing Interests in Western Europe.* Cambridge, England: Cambridge University Press.
Braybrooke, David and Charles E. Lindblom. 1963. *A Strategy of Decision.* New York, NY: The Free Press.
Bretz, Elizabeth R. 1994. "Nuclear Power: is decommissioning all that remains for the industry? or will license renewal become a reality?" *Electric World.* Volume 208, No.7 (July): 27-41.
_____. 1992. "Nuclear Power: NSSS for new plants, life extension, decommissioning." *Electric World.* Volume 206, No.8 (August): 48.

Bupp, Irvin C. and Jean-Claude Derian. 1981. *The Failed Promise of Nuclear Power: The Story of Light Water.* New York, NY: Basic Books.

Campbell, John. 1988. *Collapse of an Industry: Nuclear Power and Contradictions of U.S. Policy.* Ithaca and London: Cornell University Press.

Chubb, John E. 1983. *Interest Groups and the Bureaucracy: The Politics of Energy.* Stanford, CA: Stanford University Press.

Goldthorpe, John H. 1984. *Order and Conflict in Contemporary Capitalism.* Oxford, England: Oxford University Press.

Greenberg, Phillip A. 1993. "Dreams Die Hard." *Sierra.* Volume 79, No.6 (November/December): 102-103.

Hall, Peter. 1986. *Governing the Economy: The Politics of State Intervention In Britain and France.* New York, NY: Oxford University Press.

Hancock, M. Donald. 1989. *West Germany: The Politics of Democratic Corporatism.* Catham, NJ: Catham House Publishers.

Hatch, Michael T. 1993. "The Politics of Global Warming in Germany." Presented at the International Studies Association/West meetings. Monterey, California. 29-30 October.

_____. 1991a. "Corporatism, Pluralism and Post-Industrial Politics: The Making of Nuclear Policy in West Germany." *West European Politics.* Volume 14, No.1 (January): 73-97.

_____. 1991b. "Nuclear Power and Postindustrial Politics during the Mitterrand Era." *French Politics and Society.* Volume 9, No.2 (Spring): 14-25.

_____. 1986. *Politics and Nuclear Power: Energy Policy in Western Europe.* Lexington, KY: University of Kentucky Press.

Jacob, Gerald. 1990. *Site Unseen: The Politics of Siting a Nuclear Waste Repository.* Pittsburgh, PA: University of Pittsburgh.

Katzenstein, Peter J. 1985. *Small States in World Markets: Industrial Policy in Europe.* Ithaca, NY: Cornell University Press.

Lehmbruch, Gerhard and Philippe Schmitter (eds). 1982. *Patterns of Corporatist Policymaking.* Beverly Hills, CA: Sage Publications.

Lehmbruch, Gerhard. 1983. "Interest Intermediation in Capitalist and Socialist Systems." *International Political Science Review.* Volume 4, No.2 (April): 153-172.

Lindblom, Charles. 1977. *Politics and Markets.* New York, NY: Basic Books.

_____. 1968. *The Policy Making Process.* Englewood Cliffs, NJ: Prentice Hall.

_____. 1959. "The Science of 'Muddling Through'". *Public Administration Review.* Volume 19, No. 2 (Spring): 79-88.

Manley, John F. 1977. "Neo-Pluralism: A Class Analysis of Pluralism I and Pluralism II." *American Political Science Review.* Volume 77, No.2 (June): 368-383.

Nau, Henry R. 1974. *National Politics and International Technology.* Baltimore, MD: Johns Hopkins University Press.

Nelkin, Dorothy and Michael Pollak. 1981. *The Atom Besieged: Extraparliamentary Dissent in France and West Germany.* Cambridge, MA: MIT Press.

Novak, Viveca and Sheila Kaplan. 1988. "How the Nuclear Lobby Won Big on Capital Hill." *Common Cause Magazine.* Volume 14, No.1 (January/February): 22-23.

Rosenbaum, Walter A. 1991. *Environmental Politics and Policy.* Washington, D.C.: Congressional Quarterly Press.

_____. 1987. *Energy, Politics, and Public Policy.* Washington, D.C.: Congressional Quarterly Press.

Saumon, Dominique and Luis Puiseux. 1977. "Actors and Decisions in French Energy Politics." In Leon N. Lindberg, ed. *The Energy Syndrome.* Lexington, MA: D.C. Heath.

Schmitter, Philippe and Gerhard Lehmbruch (eds). 1979. *Trends Towards Corporatist Intermediation.* Beverly Hills, CA: Sage Publications.

Schmitter, Philippe. 1981. "Interest Intermediation and Regime Governability in Contemporary Western Europe and North America." In Suzanne Berger (ed). 1981. *Organizing*

Interests in Western Europe. Cambridge, England: Cambridge University Press.

_____. 1974. "Still the Century of Corporatism?" *Review of Politics.* Volume 36, No.1 (Summer): 85-131.

Simonnot, Philippe. 1978. *Nucleocrates.* Grenoble: Presses Universitaires de Grenoble.

Wilson, Frank L. 1987. *Interest Group Politics in France.* New York, NY: Cambridge University Press.

Chapter 8

Nuclear Politics in Soviet and Post-Soviet Europe

David R. Marples

Introduction

The Soviet Union lays claim to the development of the first civilian nuclear power station at Obninsk, a nuclear research station near Moscow, in 1954. However, the application of the atomic weapons program to the energy needs of the country did not commence in full for a further 15 years. The foundations of the Soviet nuclear power program were established in the Brezhnev period. The reasons for its development have never been explained satisfactorily. In the early 1970s, the USSR was the world's largest producer of oil — it accounted for approximately half the world's known reserves — and gas and occupied second place in the output of coal. In terms of energy resources, it could lay legitimate claims to self-sufficiency. Output of oil in the Soviet Union had risen from 31.1 million tons in 1940 to 147.9 in 1960 and 353.0 million tons in 1970. Over the same period, output of gas soared from 3.2 million to 45.3 million and 197.9 million a rise of more than 60 times. Coal production also continued to increase substantially throughout the 1960s and 1970s.

Table 1
Output of Certain of Fuels in the USSR, 1940-1970

Year	Oil[a]	Gas[b]	Coal[c]
1940	31.1	3.2	165.9
1960	147.9	45.3	509.6
1970	353.0	197.9	624.1

[a] Including gas condensate. In million of tons.
[b] In billions of cubic meters.
[c] In millions of tons.

SOURCE: Central Statistical Administration of the USSR, 1984, p. 166.

All three fuels were critical to energy output in the Soviet Union. They also appeared to meet the needs of the country adequately, in addition to providing (especially in the case of oil and gas) valuable exports that could bring in hard currency. What factors led to the change in policy and to the development of an expansive program in nuclear energy?

First, the costs of energy production had continued to rise, commencing in 1970 and increasing throughout that decade. As one energy expert has noted, the USSR failed to address this dilemma through the application of more modern techniques and increased extraction or an improvement in general management over this period. Rather, to maintain output, the planners were obliged to raise continually the share of energy in industrial investment (Gustafson, 1989: 35-36). Soviet oil output had peaked by the mid-1970s and the regime recognized the need to conserve valuable nonrenewable energy reserves. As for coal, the high-quality coking coal of the Donbass coalfield had been badly depleted after more than a century of exploitation. There were two simultaneous developments in the coal industry: a reorientation of priorities and investment from its European coalfields to those of Siberia and the Far East; and a marked decline of the famous Donetsk Basin as a source of energy supplies for Soviet electricity production.

Second, factors of geography began to play a major role in Soviet energy policy. Since 90 percent of the country's natural resources lay east of the Ural Mountains, the perennial problem henceforth would be to transport these resources to consumers in the European part of the country. Eastern Europe represented a significant factor in this equation. The countries of Eastern Europe, with the exception of Poland (which had an abundance of brown coal), were critically lacking in energy resources. The USSR had long adopted a role of energy supplier to its political allies within the countries of the Council for Mutual Economic Aid (CMEA). However, given the rise in costs and the USSR's need for hard currency, it was no longer economically expedient to provide these countries with subsidized oil and gas. An alternative had to be found.

Third, output of energy resources was subject to huge fluctuations and uncertainty. Soviet authorities sought a form of energy that could be "guaranteed" each year to meet required needs both internally and for export. Soviet industrial planning was as notable for both its precision and its general inefficiency of operation. Fourth, the centralization of Soviet industry rendered extensive debate on any form of energy production unnecessary. Critics have suggested that energy construction was subject to the whim of bureaucrats in Moscow rather than discussions at the regional level. The Ministry of Power and Electrification of the USSR was the key institution in the development of energy policy. No separate ministry existed for the various branches of energy production. A separate ministry was already in place, however, to supervise atomic weapons development, namely the secretive

Ministry of Medium Machine Building, the failings of which have been illustrated by a former Soviet scientist (G. Medvedev, 1993).

The authorities resolved therefore to meet the growing energy costs and the annual rise in the need for electricity with the development of a civilian nuclear power program. The key questions were twofold: where to locate the new installations to best meet existing needs; and what type of reactor to use. In the case of the former question, logic suggested that the stations would be built for the most part in the western regions of the country, particularly in key industrial areas and border regions (those stations that were to produce electricity for export), and as close as possible to major cities. The second issue was to pose far more serious problems for the Soviet authorities.

The Prototypes: Beloyarsk, Leningrad and Novovoronezh Nuclear Power Stations

The original civilian nuclear reactor developed at Obninsk was of the channel type based on a graphite moderator. It was thus deemed logical to conduct research into this type of reactor. The Beloyarsk station near the Caspian Sea was based on two tiny reactors with a total capacity of 300 megawatts, but it represented the pioneer project for the channel-type reactors. They were brought into service in 1963 and 1967, and the former was the first operative civilian reactor in the country. In 1965, the authorities commissioned the construction of graphite moderated reactors based on channel structure or, to use the Russian acronym, an RBMK, close to the city of Leningrad, at Sosnoviy Bor. The task was entrusted to the Kurchatov Institute of Atomic Energy and the Scientific Research and Construction Institute of Electro-Technology, which were ordered to synthesize their current experience to develop this new reactor (*Leningradskaya AES*, 1984: 28-29). It came into operation in 1973, and reached its maximum size of 4,000 megawatts (based on four 1,000-megawatt reactors) in 1981. Each reactor had some 1,690 channels.

The RBMK was developed as an exclusively Soviet type of reactor. Economically it was regarded as an efficient reactor and it could be refueled on-line. Identical RBMK-1000 stations were subsequently constructed at Kursk[1], Chernobyl[2], and Smolensk.[3] In 1983, an RBMK-1500 station at Ignalina in Lithuania came into service, which remained the largest capacity reactor in the Soviet Union until the dissolution of the USSR in 1991.[4] A further RBMK station was planned at Kostroma in central Russia. Despite such rapid expansion, the RBMK was a deeply flawed reactor that became unstable if operated at low power. It has been speculated that the prime defect lay in the construction of the control rods, which could not be inserted into the reactor core at an adequate depth.[5] It was also a design that did not provide for a large-scale accident taking place above the reactor. It was not built with the protective dome or covering common to western reactors. Altogether, according to scientists at the Kurchatov Institute of Atomic Energy, the reactor was designed with at least 32 significant defects, rendering the first generation of this reactor group among the most dangerous reactors operating in the world.[6]

---

[1]   Based on four reactors brought into service in 1976, 1978, 1983, and 1985. Six reactors were planned.

[2]   Based on four reactors with a fifth and sixth under construction in 1986. They were brought into service in 1977, 1978, 1981, and 1983.

[3]   Two reactors had been completed by the time of the Chernobyl disaster, and were brought on line in 1982 and 1985. A third and fourth were at the construction stage.

[4]   The Ignalina station began operations in 1983.

[5]   See for example, Victor G. Snell's Introduction to Marples, 1988.

[6]   In the wake of perestroika, a variety of Soviet documents were shown to the West. In 1992, James Billington at the U.S. Library of Congress organized a

A second reactor was developed concomitantly with the RBMK, namely, the water-water pressurized reactor, or VVER. Manufactured at the Volgodonsk plant in the Rostov region, the pioneer station for this type of reactor was the Novovoronezh station, which first came into operation with a VVER-210 reactor in 1964, and two VVER-440 reactors in 1971 and 1972. This reactor was developed on a broad scale as further stations were established on the Kola peninsula (1973), Metzamor in Armenia (1976), Rivne and Mykolaiv in Ukraine (the first reactors coming on line in 1980 and 1982 respectively), Kalinin (1984), Zaporizhzhya (Ukraine, 1984) and Balakovo (1985) (Petrosyants, 1985: 22). The Novovoronezh-5 reactor saw the expansion of capacity size to 1,000 megawatts in 1980, and thereafter the VVER-1000 became the most common type of Soviet reactor and the basis for future expansion.

Nuclear power at this same time became a focus of cooperation between the Soviet Union and its East European satellites. The VVER-type reactor was manufactured also in Czechoslovakia and installed there and at various other sites in Eastern Europe: Bulgaria (the Kozloduy plant); Hungary (the Paks station); and East Germany. The reactor type was also the model for a station in Finland, and was exported to Vietnam and Mongolia. Some East European countries opted for western-designed reactors. Yugoslavia's first reactor was designed by Westinghouse; while the Romanian station at Cernavoda was to be based on the Canadian CANDU (the first reactor was ready to come into service only in 1995). Energy-hungry countries such as Poland, Hungary and Romania were able to tap into the so-called MIR grid that provided energy from Soviet stations, mainly based

---

display of a number of formerly secret documents. They included a KGB top secret document which made reference to the above information released by the Kurchatov Institute.

in Ukraine. Both South Ukraine and Khmelynytsky plants in Ukraine were manufactured with investment and the use of manpower from the countries concerned: Romania and Bulgaria in the former case; Poland in the latter (Marples, 1987: 51-70).

A key period in the growth of the Soviet nuclear industry was that of the Eleventh Five-Year Plan (1981-1985), which made prognostications to the year 2000, at which time the industry was to have accounted for 30% of the electric power produced in the Soviet Union, and over 60% of that in border republics such as Ukraine and Lithuania. In 1986, it is estimated that some 36,000 megawatts of capacity were at the planning and construction stage, based predominantly on the VVER model. A new series of this reactor type was under construction with new "energy blocks" or additional reactors at the following sites: Kalinin, South Ukraine (Mykolaiv), Zaporozhzhya, Balakovo, Rivne, Khmelnytsky (Ukraine), Crimea and Rostov; with planning work under way for the commencement of construction of the Bashkir and Tatar nuclear power stations (*Energetika SSSR*, 1987: 169). Altogether 45,400 megawatts of capacity were under construction by the spring of 1986.[7] Nuclear power as a percentage of electricity production at that time was 10.5% compared to 5.6% in the mid-1970s.

This ambitious program was brought to a sudden halt by the world's worst nuclear disaster at Chernobyl on April 26, 1986. Since the Chernobyl disaster has been researched in depth by a number of scholars, I will provide here only an overview of the

---

[7]   While accidents were not reported at Soviet nuclear power stations, there is ample evidence of construction problems at the various sites, including in one case the collapse of a reactor building on its foundations. I have chronicled such dilemmas in Marples, 1987: 79-90.

main events and consequences from the perspective of almost a decade.[8]

The Chernobyl Disaster, 1986-1995

The explosion at the fourth Chernobyl reactor released over 50 million curies of radioactivity into the earth's atmosphere, encompassing between 3.5 and 10 percent of the reactor core. At least 3.5 million people within the former Soviet Union were exposed to high-level radiation. The state authorities reacted in habitual fashion immediately after the accident by refusing to divulge information to the public for some 40 hours. By May 2, the Soviet authorities had taken control at the accident scene and organized the evacuation of a zone with a 30-kilometer (18.6 mile) radius around the damaged reactor. The bulk of this zone lay within the territory of the Ukrainian Soviet Socialist Republic (SSR) with approximately 25% within the Byelorussian SSR (Belarus). The releases of radiation continued until May 10, after which a makeshift covering served to prevent further leakage. A concrete shell was constructed over the reactor and completed later in the year.

In August 1986, a Soviet delegation presented a report on the alleged causes of the accident to the IAEA in Vienna, which laid the blame on human error, and in particular the operators who authorized the dismantling of seven safety devices and the conducting of an experiment to determine how long spinning turbines would enable the use of safety equipment in the event of a shutdown, before an emergency power generator came into operation. The arrival of the Soviet delegation in Vienna might be

---

[8]   The following remarks are based on several studies of Chernobyl, including the author's. See, for example, Z. Medvedev, 1990; Marples 1987 and 1988; Kovalenko and Risovannyy, 1989; Ignatenko, 1989; US Nuclear Regulatory Commission, 1987.

perceived as a concession on the part of the Soviet authorities which had not even allowed IAEA representatives on Soviet territory prior to 1985. As a result, there was general praise for the "openness" of the Soviet report. This presentation, however, represented at best a half-truth, and the flaws in the design of the RBMK reactor were subsequently revealed posthumously by the leader of the Soviet delegation to the IAEA, Valeriy Legasov, who committed suicide on the second anniversary of the accident.

Nonetheless, the western scientific community generally accepted that the Soviet authorities acted in good faith after Chernobyl. Though the 30-kilometer zone encompassed only a fraction of the area subjected to radioactive fallout from the nuclear plant, few questions were raised about the viability of evacuating such a small area, or about the lack of precautions being adopted in agriculture, nutrition and child rearing in the neighboring areas. In reality, the secrecy in which the major events of Chernobyl were maintained by the authorities was equal to that for any previous accident, including the explosion of a nuclear waste dump at Kyshtym in the Urals in 1957. All information on health effects, for example, was strictly classified. In early 1989 when the first maps of the fallout area were published in the all-Union and republican press, they caused widespread panic among the population which believed, with some justice, that it was the victim of a large-scale deception.

While a national all-Union program was in place by 1989, which provided compensation for Chernobyl victims and organized subsequent evacuations of populations in highly irradiated zones,[9]

---

[9] By 1989, the main danger to the population arose from the soil rather than the air. In particular, Cesium-137, which has a half-life of 30 years, was used as the yardstick to determine the need for evacuations. At levels of over 40 curies per square kilometer density in the soil, evacuations were ordered immediately. Next, people in areas with 15-40 curies were to be resettled. Areas of 5-15 curies were designated as zones of "secondary evacuation" while by 1989-90 all areas with more than one curie of cesium in the soil were labelled zones of periodic control.

the authorities in Moscow began to lose control over the Chernobyl problems as a result of self-assertion at the republican level. In 1991, both Ukraine and Belarus adopted their own state programs, and lowered the official tolerance level per individual to 1 millisievert (mSv) per year from the original 5 mSv. Belarus suffered approximately 70% of the high-level fallout in the former Soviet Union and about one-fifth of the republic (embracing 2.2 million people out of a population of 10.3 million) was affected.

In the long term, the evacuations that were conducted have proven to be generally unsatisfactory to those concerned. First, the new locations rarely met the needs of the resettlers. Often there were no jobs to go to. The new homes were constructed hastily and were often lacking in basic amenities. Although there was initially a strong desire among residents to leave the contaminated territory, after the first batch of settlers had been moved, those remaining behind were more reluctant to follow suit (most likely because of the much publicized complaints of the evacuees about their new homes). Second, the evacuations appear to have caused great stress among families, particularly among the elderly who had resided in their native villages since birth. Third, a large majority of potential evacuees used the opportunity to request relocation to a major city, particularly Minsk (in the case of Belarus) and Kyiv (in the case of Ukraine). Fourth, the evacuations were uneven. Frequently there was no correlation between the amount of radiation being measured in the soil and that entering the human organisms. Thus in areas of very-low level contamination (less than one curie per square kilometer of cesium), morbidity levels were often higher than in areas with relatively high levels.

As for the medical impact of Chernobyl, the question has unfortunately been shrouded in controversy from the outset, largely

---

The key factor was the"acquired equivalent dose of irradiation to the body" that was likely to arise from the consumption of contaminated food.

because of a deep rift between the scientific community (represented both by the Soviet ministries of health and nuclear power on the one hand, and the IAEA on the other) and the general public and environmental organizations in the Soviet republics on the other. A study commissioned by the IAEA in 1990, which was published in the spring of 1991, failed to identify any significant health effects during a survey of 13 villages (6 control villages and 7 in the contaminated areas) that could be unequivocally attributed to Chernobyl.[10] This preliminary survey — it was no more than that — was greeted with fury by politicians and scientists alike in the republics concerned.

By the end of 1994, however, it was possible to make several deductions about the medical impact of Chernobyl. First, casualty rates and sicknesses were notably high among the cleanup crews who had worked on removing radioactive debris from the roof of the destroyed reactor and other hazardous tasks exposing them to very high levels of irradiation in the summer months of 1986. While official Ukrainian figures which numbered mortalities in the tens of thousands have not been possible to corroborate,[11] there is evidence that the military reservists who entered the zone in May 1986 have suffered substantial casualties. In addition, the

---

[10] International Advisory Committee, 1991.

[11] In an interview with the author held in September 20, 1990, a former official in the town of Chernobyl who worked for the Pripyat Industrial and Research Association stated that a figure of 5,000 dead, which was being circulated by inter alia members of the Green World ecological association, was "not unrealistic" (Risovanny, 1992, p. 148). More recent figures from the Ukrainian Ministry of Health suggest casualty figures of over 125,000 in the period 1988-1994 (*The Ukrainian Weekly*, April 30, 1995). However, these figures have yet to be validated, i.e., there has been no list of those alleged to have died and even were such a list to appear, one would still have to determine how many of these deaths could be attributed directly to work in the Chernobyl zone and radiation from the disaster of 1986.

"liquidators" (to use the Russian term) have suffered from digestive problems, diseases of the blood, and a great variety of ailments since 1986. Many have died from heart attacks that may have resulted from the stress to which they were subjected in the period of decontamination (Risovanny, 1992).

The number of leukemias among the evacuees, cleanup workers and affected population, however, has not risen significantly. Though the number per se has increased since 1986, it is difficult to determine a precise cause. The rise in numbers remains within the European average and the total annual figure for leukemias in 1986-1991 remained significantly lower than the peak years of the 1970s. On the other hand, the rise in thyroid cancers among children is alarming, and largely confined to the contaminated zones (the Homel and Brest regions of Belarus; the Kyiv, Zhytomyr and Chernihiv regions of Ukraine, and the Bryansk Oblast in southern Russia). In the single worst affected area, Homel Oblast in Belarus, the level of thyroid gland cancer among children is today 100 times higher than before the accident. Altogether 333 Belarusian children were suffering from this highly aggressive form of cancer by the end of 1994 (Demidchik, 1995).[12]

Medical experts are in general agreement that the chief cause of this phenomenon is radioactive iodine released from the Chernobyl reactor immediately after the disaster (the radionuclide has a half-life of only eight days). Iodine fallout encompassed about 80% of Belarusian territory, for example, but the heaviest fallout was in the border regions of Homel and the Polessye marshes. This area is known for the deficiency of iodine in the soil — levels of goitre and thyroiditis are particularly high — and hence the thyroid

---

[12]   The above information was ascertained by the author during seven visits to Belarus in the period 1992-1995 and interviews with medical specialists. A more detailed study is in press, David R. Marples, *Belarus: Soviet Rule and Nuclear Catastrophe*, forthcoming, The Macmillan Press, 1996.

glands of the children in particular were especially susceptible to the radioactive iodine in the atmosphere.

Generally the levels of morbidity in Ukraine and Belarus today are significantly higher than in 1985, although it is rarely easy to determine the direct cause of what has been termed a health crisis. Local doctors sometimes attribute the rise in diseases such as diabetes mellitus among children to the increased radiation levels brought about by Chernobyl. Logically, one must take into account other factors such as lifestyle, industrial pollution, a generally poor diet and the fall in the standard of living that has occurred in the period 1991-95.

Chernobyl also sparked a powerful antinuclear movement across the Soviet Union, epitomized by the Green World ecological association in Ukraine, which organized pickets and demonstrations at a variety of nuclear power stations and other industrial establishments considered to be ecologically dangerous.[13] This movement was most effective in the period 1988-91 and resulted in the abandonment, stoppage of work or postponement of some 60 planned reactors. Among the stated reasons for such curtailments were proximity to large cities, a lack of waste disposal sites, the presence of reactors in zones of significant seismic activity, and above all local anger at the centralization of the nuclear power program in Moscow (Marples, 1991: 133-144). By 1991 it appeared that the future of this industry in the Soviet Union was a limited one.

---

[13]　Interestingly, the leader of this movement, Dr. Yuriy Shcherbak, became Minister of the Environment in 1991. Subsequently he accepted the position of Ukrainian ambassador to Israel, effectively beheading the movement. Today this highly respected politician and doctor is the Ukrainian ambassador to the United States.

The Recovery of the Atomic Energy Industry, 1992-1995

The recovery of the nuclear industry in the countries of the Commonwealth of Independent States (CIS) has been quite remarkable. In late 1992, spokespersons for the Russian atomic power industry (Rosenergoatom) announced an ambitious new program for the expansion of nuclear energy in Russia, with emphasis on plant construction in the remote areas of the federation. The program — though it lacked adequate funding — represented a symbolic resurgence of the nuclear power lobby in Russia, the offshoot of the former Ministry of Nuclear Power and Engineering, almost seven years after the Chernobyl disaster. The scientists anticipated that nuclear power could resolve Russia's current energy problems in the oil and gasfields. They also noted that three reactors had been at an advanced state of construction in 1986 (Balakovo-2; Kalinin-3; and Kursk-5) and could quickly be brought into operation. New reactors for existing units were designated for the Kola peninsula, Sosnoviy Bor (St. Petersburg), Khabarovsk and Kostroma, and new units were also anticipated at Novovoronezh, the oldest and original location of the VVER reactor in Russia.[14]

The goal of the new program was declared to be to raising capacity at Russian atomic power stations from 20,000 megawatts to 37,000 megawatts by the year 2010. The program thus appeared to be a modest one, but it nonetheless signified a radical change of view in ruling circles about the viability of nuclear energy as a leading power source for the future. A decline in the production of oil and gas in recent years is likely to continue and will perhaps be compensated by a resurgent nuclear power industry. In turn, those neighboring states in the CIS that have long been dependent upon

---

[14]   News of the new program was announced at a press conference in Moscow on January 20, 1993.

Russian energy supplies — Ukraine and Belarus, in particular — have also been obliged to address the question of, in the former case, the further development of nuclear energy and, in the latter case, the creation of a domestic nuclear energy program from the beginning.

Facing economic crises, the newly independent states of Ukraine, Belarus, and Armenia in 1992-93 began to address the question of nuclear power development. In the case of the Ukraine and Belarus, authorities faced a huge psychological barrier in that the population, having suffered the brunt of the radioactive fallout and the medical repercussions of Chernobyl, was very wary of any nuclear power program. In April 1992 and 1994, the most powerful charitable organization in Belarus, the Belarusian Charitable Fund for the Children of Chernobyl, sponsored international conferences in Minsk under the general title "The World After Chernobyl." In both cases, an avowed aim was to provide alternatives to nuclear power development in the republic. For the planning authorities, however, the immediate concern was the state's inability to cover the costs of imports of oil and gas in hard currency payments. In Armenia, on the other hand, the chief concern was the location of the Metzamor station, which had been shut down in March 1989 following the major earthquake in that country three months earlier. The Armenians signed an agreement with Russia in March 1994 to provide nuclear fuel for the proposed restarting of the Metzamor station (*Interfax*, May 5, 1994).

Ukraine and Belarus appeared to follow the lead of Russia on the nuclear power issue. However, the ultimate decision has remained that of the individual state leaders. In Ukraine's case, nuclear energy seemed a logical alternative to traditional energy fuels. Nuclear energy was already responsible for over 30% of total electricity output and hence could hardly be abandoned outright without drastic repercussions on the population. Ukraine in 1992 had five existing stations with a total of 14 reactors. Moreover, like Russia, it possessed several reactor complexes at an advanced stage of construction: Zaporizhzhya-6; Khmelnytsky-2; and Rivne-5. They had been halted by a 1990 moratorium placed on further reactor building and operation by the Ukrainian parliament.[15]

The Ukrainian stations could thus be brought into service within a period of 6-18 months and the consensus within energy circles was that they could make a significant impact upon the Ukrainian economy. In 1992, for example, the Zaporizhzhyan station alone (with five reactors operational) produced 31.5 billion kilowatt hours of electricity, or 13.3% of total output in Ukraine, and the equivalent of 12 million tons of coal, or 7.6 million tons of oil expended for the same output of electricity at thermal power stations (*Pravda Ukrainy*, April 24, 1993, p. 2). One Ukrainian scientist declared that a single unit at the Chernobyl station — then under review (see below) — could supply enough electricity to meet the needs of the entire city of Kyiv, with a population of 2.5 million, with electricity (*Vechirniy Kyiv*, April 29, 1993, p. 2).

Turning to the Republic of Belarus, the only reactor under construction at the time of Chernobyl — a nuclear power and heating station some 20 miles east of Minsk — was abandoned in 1988 after widespread public protests. The location of the republic, however, has rendered it very susceptible to the effects of any nuclear accident involving Soviet manufactured RBMK reactors. On the northwest border is located the Ignalina station, an RBMK-1500, on Lithuanian territory. Some 40 miles to the east is the RBMK-1000 near Smolensk; and 15 miles to the south of the southern border is the Chernobyl station itself. Skeptics have

---

[15]   The other stations operating in Ukraine are Chernobyl itself and South Ukraine (Mykolaiv). The latter is part of a vast energy complex which includes two hydroelectric stations. The moratorium has stalled not only the fourth reactor at the Mykolaiv site, but also work on the hydroelectric stations. See *Robitnycha hazeta*, December 18, 1992, p. 3.

therefore argued that Belarus might profitably pursue its own program since the dangers of an accident from a domestic station could hardly surpass those from the stations along the borders of the country. As Russian oil supplies to Belarus have varied according to the timeliness of payment, Belarus' nuclear power lobby has become a significant one. Four possible sites have been selected for the country's first nuclear power station, while scientists have argued that the cost of importing nuclear fuel from Russia is considerably less than the price for imported Russian oil and gas. Nevertheless, opposition to such plans remains significant.

In Ukraine also, the pro-nuclear lobby was able to marshal some convincing arguments in favor of a renewed program. By 1993, the government was under strong pressure to lift its moratorium on new construction. For example, the Ukrainian Supreme Soviet received a petition early in 1993 from the city of Enerhodar, the city that houses the workers and operatives of the Zaporizhzhya station, to end the moratorium forthwith. One petitioner noted that France, a country of comparable size in area and population to Ukraine, had 56 reactors in service compared to Ukraine's 14. Other relatively small countries were also pursuing the nuclear option: Japan had 43 reactors in an area almost half that of Ukraine; South Korea planned to bring into service 18 new reactors at its atomic power stations by the year 2010 (*Pravda Ukrainy*, April 29, 1993, p. 1).

The Debate over the Chernobyl Station, 1993

Nineteen ninety-three was the pivotal year for the Ukrainian nuclear power industry. In this year, a dramatic reversal of policy occurred as a result of the efforts of a strong pro-nuclear lobby, and Ukraine's continuing energy difficulties. When the decommissioning of the Chernobyl plant began in 1993, with the planned removal of the second reactor from the grid (it had been shut down after a fire in 1991)[16], there was a chorus of protests at the highest level, advocating delay and even abandonment of the schedule to take the reactor out of service. A government official pointed out that once fuel had been removed from the reactor core, the process of decommissioning would be almost impossible to reverse. In his view, money spent on making Chernobyl more reliable and safe would be wasted if the plant was closed down. About 270 million rubles (in 1984 prices) had been expended on improvements, almost 25% of which was devoted to work on the reactor itself. In the official's view, these changes had rendered Chernobyl one of the safest of the former Soviet RBMKs (*Pravda Ukrainy*, April 29, 1993, p. 3).

The above comments found further backing from Bronislav Pschennikov, a candidate of technical sciences from Ukraine. Pschennikov acknowledged that Chernobyl in retrospect should not have been brought back into service in 1986.[17] That decision was basically a political one to display to the world that the consequences of Chernobyl had not been as substantial as first feared. It was a decision that exposed thousands of workers to very high levels of irradiation. Nevertheless, he added, the Chernobyl reactors had been the subject of serious study. Since 1986, several

---

[16]   According to the director of the then Ukrainian State Committee for Nuclear and Radiation Safety (now part of the Ministry for the Protection of the Environment and Nuclear Safety), Nikolay Shteinberg, although the Ukrainian parliament had resolved to close unit two permanently on October 22, 1991, this decision was also reversed on October 22, 1993, when the moratorium was lifted. Steinberg, 1994-1995: 45.

[17]   Chernobyl's Reactor No. 1 was back on line only five months after the major accident, on October 1, 1986. The second reactor followed a month later. Reactor 3, which posed more serious problems in light of its adjoinment to the destroyed fourth reactor, was returned to the grid only in December 1987. See Marples, 1988.

improvements had been introduced: the reactor was provided with a higher enrichment of uranium; its shutdown time was reduced significantly; and new laws were issued preventing experiments without the presence of the plant director and chief engineer. Radiation levels had also fallen each year and by 1993 were at only 20% of the permissible norm. According to the principal builder of the station and its chief scientist, the first reactor unit could operate at Chernobyl until 1997, and the third unit until 2001. Pschennikov also noted that developed countries such as Britain continued to operate stations that were much older than Chernobyl, some of which dated to the 1950s. The then-president Leonid Kravchuk was noncommittal in the face of international pressure to close the plant, pointing out that an international research center for the nuclear industry could be established at the station, using the city of Slavutych, 40 miles to the east in Chernihiv Oblast as a base for the workers (*Vechirniy Kyiv*, April 29, 1993, p. 2).[18]

Other observers in Ukraine noted in the spring of 1993 that with enhanced safety and ecological awareness, Chernobyl could continue to operate and benefit the Ukrainian economy (*Pravda Ukrainy*, April 22, 1993, p. 1). Ukraine has, however, suffered the loss of several thousand nuclear experts since gaining its independence in August 1991, mainly through the decision of qualified Russian technicians to return to Russia (*News From Ukraine*, No. 15, April 1993, p. 2). This deficit of well qualified specialists has been compensated partially by the convocation of a

---

[18]   Slavutych was constructed as a center for Chernobyl workers to replace the evacuated city of Pripyat, which was declared uninhabitable after the Chernobyl disaster. For an account of the radiation situation there on the eve of its construction, see the secret protocol of the USSR Council of Ministers, "Concerning the site for the construction of a new settlement for the permanent habitation of the personnel of the Chernobyl nuclear power plant and their family members." Secret Protocol, circa August 13, 1986, cited in Yaroshinskaya, 1992: 466-467.

number of high-level international conferences and meetings in the republic that was the location of the world's worst nuclear accident. In the spring of 1993, for example, a high level conference of the International Atomic Energy Agency (IAEA) was held at Enerhodar, in which nuclear experts from Sweden, the United States, Spain and Ukraine participated (*Pravda Ukrainy*, April 17, 1993, p. 2). There the IAEA specialists had an opportunity to examine the construction of the new, sixth reactor that, once on line, would render Zaporizhzhya the largest nuclear plant in the world.   This conference followed a "successful" meeting between Kravchuk and IAEA Director Hans Blix in December 1992, during which the Ukrainian president declared that his country could not survive without nuclear energy, a clear indication that the moratorium was about to be lifted (*Demokratychna Ukraina*, December 22, 1992, p. 1).

A further reflection of the increasing success of the pro-nuclear campaign was the government's decision to privatize the Rivne nuclear power plant and free it from the jurisdiction of the State Atomic Inspection Committee. The latter's role was reduced to supervision over questions of safety rather than the daily operations of the plant (*Robitnycha hazeta*, April 30, 1993, p. 1). The Fall 1993 session of the Ukrainian parliament lifted the moratorium and extended the lifespan of the remaining reactors at Chernobyl to their "natural limits" of 25-30 years. The matter did not, however, end here. An IAEA delegation visited the Chernobyl site in the spring of 1994 and declared the reactor too dangerous to remain in service. The U.S. Nuclear Regulatory Commission agreed with this verdict and its director suggested that international organizations and financial institutions should assist Ukraine to commission its three almost completed reactors which would then compensate for the loss of the Chernobyl station (Selin, 1994/95: 10). In the spring of 1995, after strong international pressure, Ukraine's new president, Leonid Kuchma, informed representatives of the European Union and G-7 countries that he would halt operations at the Chernobyl station by the year 2000 (Kolomayets 1995: 1).

Conclusion

Unfortunately, the Chernobyl story is far from over. The covering over the damaged reactor will have to be replaced within the next 10-15 years. Moreover, even the proposed closure of the nuclear station may be dependent on an improvement of Ukraine's energy situation. Elsewhere in the CIS countries, nuclear power plants are notably lacking in safety procedures and since 1991 accidents have been frequent, including a fatality at the Zaporizhzhya station in 1993 (UPI, May 22, 1993).[19] The new Russian program was announced before funds had been assigned. At the time of writing it appeared to be inconceivable that the Russian government would allocate such funding. Hence both Russia and Ukraine remain very much dependent on the charity of international organizations both to fund a future program and to render it safe.

In the meantime, the two Slavic states and other members of the CIS have demonstrated an alarming tendency to "muddle through" and maintain present operations at nuclear power plants with reduced levels of safety. Hence while the RBMK has now been almost universally condemned as a dangerous reactor, and attention has periodically been focused on the need to stop operations at Chernobyl, very little has been said about the remaining five stations still in operation, including the giant Ignalina station, once bitterly opposed by the Lithuanian Sajudis, but now responsible for some 80% of that country's electricity

---

[19] The author has discussed recent accidents and the safety question in general in Marples, 1994.

supply. The question of continued operations at such plants must be discussed by the IAEA and other international organizations if the world is to avoid a major nuclear accident in the immediate future. Despite the world's worst nuclear accident, nuclear power in the territories of the former Soviet Union today is less safe than it was in 1986.

References

Central Statistical Administration of the USSR. 1985. *Narodnoe khozyaystvo SSSR v 1984g.* Moscow.

Demidchik,E.P. 1995. "Thyroid Cancer After Chernobyl." Unpublished Statistics. Minsk.

Energetika SSSR v 1986-1990 godakh. 1987. Ed. A.A. Troitskiy. Moscow.

Gustafson, Thane. 1989. *Crisis amid Plenty: The Politics of Soviet Energy under Brezhnev and Gorbachev.* Princeton, NJ.

Ignatenko, E.I. ed. 1989. *Chernobyl: sobytiya I uroki. Voprosy I otvety.* Moscow.

International Advisory Committee. 1991. *The International Chernobyl Project.* Vienna: International Atomic Energy Agency.

Kolomayets, Marta. 1995. "Ukraine to shut down Chernobyl by 2000." *The Ukrainian Weekly.* April 16: 1-4.

Kovalenko, Aleksandr and Yuriy Risovannyy. 1989. *'Chernobyl' — kakim ego uvidel mir.* Kiev.

Leningradskaya AES. 1984. Ed. V.M. Babanin et al. Leningrad.

Marples, David. 1994. "Nuclear Power in the CIS: A Reappraisal." *RFE/RL Research Report.* Volume 3, No.22 (3 June): 21-26.

_____. 1991. "The Greening of Ukraine: Ecology and the Emergence of *Zelenyi svit, 1986-1990.*" In Judith B. Sedaitis and Jim Butterfield (eds). *Perestroika From Below: Social Movements in the Soviet Union.* Boulder, CO: The Westview Press.

_____. 1988. *The Social Impact of the Chernobyl Disaster.* London: The Macmillan Press.

_____. 1987. *Chernobyl and Nuclear Power in the USSR.* London: The Macmillan Press.

Medvedev, Grigorii. 1993. *No Breathing Room: The Aftermath of Chernobyl.* New York, NY: Basic Books.

Medvedev, Zhores. 1990. *The Legacy of Chernobyl.* New York, NY: W.W. Norton.

Petrosyants, A.M., ed. 1987. Atomnaya nauka I tekhnika SSSR. Moscow.

Risovanny, Yu. 1992. "An Insider's View of Chernobyl." In Solchanyk, Roman, ed. *Ukraine From Chernobyl to Sovereignty.* New York, NY: St. Martin's Press.

Selin, Ivan. 1994/1995. "Opening Remarks: Symposium on Nuclear Safety in the Soviet Union." *CIS Environmental Watch.* Number 7, Fall /Winter: 5-17.

Steinberg, Nicolai. 1994/1995. "Nuclear Safety and the Regulatory Processes in Ukraine." *CIS Environmental Watch.* Number 7, Fall/Winter: 45-51.

U.S. Nuclear Regulatory Commission. 1987. *Report on the Accident at the Chernobyl Nuclear Power Station.* Washington, D.C.: NUREG 1250.

Yaroshinskaya, Alla. 1992. *'Chernobyl': sovershenno sekretno.* Moscow.

Chapter 9

The Asian Atom:
Hard-Path Nuclearization in East Asia

Jong-dall Kim and John Byrne

Introduction

Nuclear power has long been promoted in the West as a limitless energy resource. This history of promotion is once again being repeated, this time in one of the world's economically fastest growing regions, East Asia. Unlike the countries of Europe and North America, where orders for new nuclear plants have largely been suspended, the countries of East Asia (i.e., Japan, South Korea, North Korea, Taiwan, and China) have or will soon be investing heavily in nuclear technology.

A primary impetus for the movement into nuclear energy has been the rapid economic and energy growth being experienced in East Asia. While the rest of the world experienced an overall rate of energy growth of 3 percent during the 1980s, Asia was expanding its energy usage by 5 percent per annum and, in countries like South Korea, the annual rate of growth exceeded 7 percent (Byrne et al, 1992). Given the paucity of indigenous energy resources available to many of these countries (in particular, Japan, North Korea, South Korea, and Taiwan) nuclear power is seen by many in the region as the most sensible form of energy investment. However, despite the significant role that nuclear power seems destined to play in the region's energy, and indeed, social, future, remarkably little research has been undertaken concerning the socio-political aspirations which are guiding this development path.

The countries of East Asia display a great diversity of social and political forms. While China remains the last of the Communist giants, only recently opening itself up to quasi-capitalist market organizations, Taiwan, Japan, and South Korea, have, at least since World War II, been in the forefront of state-directed capitalist development. North Korea remains perhaps the world's most secretive country with a highly centralized planned economy. Yet, despite the great diversity of political and economic regimes, countries in the region have created remarkably similar nuclear technocracies.

The chapter opens with a general overview of nuclear developments in East Asia. Attention then turns to the Japanese and South Korean experience as case studies in technocratic development of nuclear power in the region. The last two sections discuss the social consequences of nuclearization of national energy policies and the variety of problems which accompany the spread of the technology in the region. The increasing level of nuclear tension in the region will also be examined, primarily as a result of North Korea's integration into the nuclear technocratic order in exchange for promises of non-proliferation.

Nuclear Developments in East Asia

East Asia is rapidly becoming the world's largest producer of nuclear-generated electrical energy. The intensive development of the technology in the region can be explained both as the result of each nation's desire to create technocratic competency and a shared perception that the technology represents the most economical response to energy demands stimulated by economic growth. This is not to say that the nuclear development paths of the countries of the region have been completely uniform. For instance, while Japan and South Korea have invested enormous financial and political resources in civilian nuclear power expertise and infrastructure since the 1950s, China and North Korea have only recently begun to develop the civilian side of the atom.

In part, this difference in emphasis can be explained by the unique situation faced by each of the nations. While Japan, South Korea and Taiwan are resource-poor countries and import virtually all of their fossil fuels, China and North Korea have abundant domestic energy resources. Only recently have the structural problems faced by China, namely significant difficulties in the transportation of fossil fuel from the interior of the country to the primary industrial and residential areas in the coastal areas, and the high sulfur content of indigenous coal supplies, led to the decision to create a nuclear power supply system. North Korea also has significant hydroelectric energy resources but lacks the capital required to construct energy facilities and has turned to nuclear power as a means of gaining capital for its energy needs from outside the country.

The historical emphasis placed on nuclear power is clearly evident in Table 1, which offers a comparison of the relative nuclear capacity of the nations of East Asia. Japan possesses the largest number of both research and power reactors in the region and is the furthest along in the planning of new reactors. South Korea's program is also very strong, following only Japan in most categories. China's commitment to the nuclear energy option is in the relatively early stages of preparation, at least compared to both South Korea and Japan. Nonetheless, China's decision to build a significant number of new plants, and its massive energy needs, makes it a likely center for rapid nuclear power development by the early part of the next century. Taiwan's nuclear power program is comparatively modest, but in terms of its importance to electrical supply, the country is quickly increasing the role of nuclear technology. Finally, North Korea is a recent entrant but with possibly substantial nuclear ambitions (civilian and military).

TABLE 1

Nuclear Reactors in East Asia

	Japan	South Korea	Taiwan	China	North Korea
Research reactors	18	3	0	8	1
Power reactors in operation	42	10	6	1	1
Power reactors under construction	10	6	0	3	2
Power reactors in planning	13	7	2	7	3

Note: Reference year for South Korea is 1995

Source: International Atomic Energy Agency. Korea Electric Power Corporation (KEPCO); and Institute of Energy Economics, Japan.

Japan

Japan argues that it has developed nuclear energy primarily in response to limited resource endowments. As a result of the combined efforts of the government and major industrial conglomerates, nuclear energy now supplies some 32,044 megawatts (MW) of electricity. This represents about 10% of Japans's domestic energy needs or about 24% of total guaranteed electricity (Japan Electric Power Information Center, 1994: 10). Nuclear's contribution will increase significantly in the coming years, with the share of nuclear power generation in total generated output being expected to rise to 32% (300 TWh) by the year 2000 and to 35% (380 TWh) by 2010, with further increases expected throughout the next century: 50% (680 TWh) by 2030, 65% (960 TWh) by 2050, and 65% (1,250 TWh) by 2100 (Kato, 1993: 32). The ten reactors currently under construction will increase total nuclear capacity to 45,600 MW by 2000 with additional planned reactors increasing total net capacity to 75,000 by 2010. This investment will place Japan third on the worlds' list of leading nuclear generators, following only the United States and France.

Japan has been a leader in the development of indigenous nuclear capacity. The stated rationale for this effort has been "national energy security" including the need to protect itself against the energy shocks like those that resulted from the oil embargos of the 1970s. The country has also followed a policy of nuclear diversification. At least two important policy implications flow from the simultaneous policies of indigenization and diversification. First, the nation's planners see nuclear technology as an important export product for the next century. Thus, it is argued that nuclear power not only frees the nation from the burden of an almost total reliance upon fossil fuel imports but also serves to extend its economic reach in the region and the Pacific at large. Second, Japan operates a number of different types of power reactors, including two fast breeder reactors (FBR) at Monju and Joyo and a number of enrichment facilities and spent fuel reprocessing facilities, as well as an experimental plutonium reactor. This latter facility is particularly controversial given the technology's hazardous operational and post-operational problems. Japan expects the FBRs to play a key role in maintaining adequate baseload capacity while simultaneously reducing overall greenhouse gas emissions. Indeed, by the 21st century, a full-scale shift to FBRs is being projected.

South Korea

South Korea has fundamentally changed its energy supply structure in only three decades. The most important part of this change has been the now-heavy reliance upon nuclear technology. By the end of 1993, the nation's nuclear capacity stood at 7,616 MW or 25% of total installed electricity capacity. Nuclear power production was 58,138 GWh, or 49% of total electricity output. Faced with a rapid growth in demand that regularly exceeds 10 percent per annum, South Korea plans to bring on line 14 new reactors with a total capacity of 20,416 MW by 2006, at which time over half of the nation's electricity will be nuclear-supplied.

Unlike Japan, South Korea has focused its efforts exclusively upon pressurized-water reactors (PWR). Consequently, it expects to be fully self-reliant in PWR design, engineering and construction by the middle part of the 1990s. By 2001, the Korean Atomic Energy Research Institute (KAERI) plans to have developed the next-generation Korean reactor, essentially an advanced version of the Korean Standard Nuclear Plant (KSNP). KAERI also has a program in place to develop all major nuclear options, including liquid metal reactors (LMR), fusion reactors, advanced PWRs, and CANDUs, and fuel recycling and disposal technologies.

Taiwan

Taiwan has much in common with Japan and South Korea. As a resource-poor country driven by a dynamic export-oriented economy, Taiwan views the nuclear option as one of the principal ways to achieve energy security. However, unlike its neighbors, Taiwan has chosen not to invest heavily in an indigenous nuclear infrastructure overseen by the state. Instead, it has allowed the dominant utility, Taipower, to supervise the construction and operation of the nation's nuclear facilities.

Taiwan currently operates six reactors, with a combined capacity of 5,144 MW. Four of the reactors are General Electric boiling water reactors (BWR) and two are Westinghouse PWRs. Six additional plants are planned in the next century. The builders of these plants, all of which will be operated by Taipower, are as yet undecided.

China

China entered the civilian nuclear era in December 1991 when it inaugurated the 300 MW Qinshan power station. While China has plans for a relatively modest seven reactors, many analysts regard China as the most important potential new nuclear market in the 21st century. With economic growth averaging 9 percent per year since 1978 and 13 percent since 1992, and with a quarter of the world's population, the Chinese market for new energy sources seems virtually unlimited. While GNP and per capita GNP doubled between 1980 and 1990, its annual electricity consumption increased from 3 million kWh to 5.5 million kWh. By comparison, prior to the 1980s, per capita annual electricity use averaged less than 100 kWh in cities and just 20 kWh in rural areas. By 1990, the annual average per capita use increased to 494 kWh. Despite the nation's ability to meet this rapid increase in demand, there is nonetheless a perceived shortage of basic electricity supplies. According to one report, at least 10,000 additional MW per year will be necessary to meet potential energy shortfalls (*Nucleonics Week*, 1994: 3).

The major question facing China is the future structure of its energy supply system. As noted above, China is rich in natural resources, but its coal, oil and gas tend to be far removed from its industrial and urban areas along the Pacific coast. Its transportation infrastructure is also relatively underdeveloped, making the shipment of bulk fossil fuels both costly and time-consuming. The tendency of these economic constraints to militate towards the development of the nuclear option is reinforced by the ability of civilian planners to borrow freely from existing weapons technology. In this respect, China has copied both its international and regional rivals in the development of its nuclear program. Much like the United States and Russia, China has invested heavily in atomic weapons prior to seeking the commercialization of nuclear technology. On the other hand, as is the case in South Korea and Japan, China is seeking to indigenize its nuclear capacity. Thus, it now has plans in place to design and manufacture a standardized plant of Chinese origin (Kang, 1993).

The indigenization effort is paralleled by the world's most ambitious nuclear expansion program: 3,500 MW to be built by 2000, 50,000 MW by 2015, and 350,000 MW by 2050 (*Nucleonics Week*, 1994: 3). A modest start to this program was marked by the 1994 start-up of two 950 MW reactors at Daya Bay near Hong Kong. China is also constructing two 650 MW power plants of its own design at Qinshan; these plants will come on-line in 1998. Two additional 1,000 MW reactors of Russian design are scheduled for completion by the turn of the century (Han, 1995: 16).

North Korea

North Korea's nuclear program seems to have contributed more to the development of an atomic weapons system than adding to the country's means for generating electricity. The nation has built a large reprocessing facility; it has also succeeded in extracting weapons grade plutonium from its existing graphite reactors. However, it is not clear what has been done with these capabilities or with the plutonium. There is great suspicion of the North's intentions in the region, particularly since the experimental 5 MW reactor at Yongbyon, which is a graphite-moderated, natural uranium-fueled and gas-cooled reactor, is more appropriate for generating weapons-grade plutonium than electricity.

Recently, North Korea has been seeking assistance for the construction of two PWRs. These units would be built in exchange for the shutdown of the previously mentioned graphite reactors. It appears quite likely that South Korea and Japan, operating under U.S. and international supervision, will build and finance the PWRs.

Technocratic Centralization in East Asia

The roots of the Asian atom can be traced to the United States and its European allies. When President Eisenhower announced the "Atoms for Peace" program in 1953, he proposed that humanity "utilize the destructive power of the atom to serve the peaceful pursuits" of mankind and to provide "abundant electrical energy" (Hilgartner et al, 1983: 41-42). This message appealed to developing countries seeking a level of technological and electricity independence, as well as to technically developed countries whose military-based nuclear "technostructures" were searching for a peaceful nuclear mission and a market for their inventions (Winner, 1977). However, nuclear power technology could not be exported overseas unless developing countries acquired the knowledge and appreciation of nuclear technology. It was in this global political economy of nuclear technology promotion that the U.S. and IAEA offered basic nuclear information to Japan and South Korea, leading to the formation of the first corps of domestic nuclear scientists and experts in the 1950s. Westinghouse and General Electric in Japan and Westinghouse in South Korea served as the conduit for these promotional missions.

Japan entered the nuclear power era earlier than any other East Asian country, with its development proceeding steadily and systemically during the 1960s and 1970s. The central government developed its first nuclear plan at the end of 1962. Under the plan, the three largest Japanese utilities, Tokyo, Kansai and Chubu, were assigned the tasks of technical training and were held responsible for developing long-term plans for the indigenization of nuclear power technology. Other utilities were to build reactors only after the economies and reliability of nuclear power were established.

Private sector *zaibatsu* (also called *keiretsu*) were mobilized to assist in the goal of building a Japanese power system during the 1960s and 1970s. Following Westinghouse's construction of Mihama-1 PWR in 1970s, the Japanese conglomerate Mitsubishi completed Mihama-2 PWR in 1972 and subsequent power plants. Along with Hitachi and Toshiba, Mitsubishi, in alliance with state planners and foreign companies such as Westinghouse, GE and GEC (UK), constructed some 47 other plants. By concentrating the technology in the development plans a very few companies, Japan was able to indigenize nuclear power technology and to export it to other developing countries.

A very similar process of indigenization has taken place in South Korea. Since the Korean War (1950-1953), the country has also undergone a period of rapid urbanization and industrialization. To meet the energy needs of its growing economy, South Korea created a large-scale, highly centralized system of energy production. A single public utility, the Korea Electric Power Corporation (KEPCO) was organized and given responsibility for the construction of power plants, the generation and distribution of electricity, as well as planning and finance for future energy needs. Foreign companies such as Westinghouse and Framatome were major players in the early stages. Private sector *chaebols*[1] were enlisted to assist in the goal of building a South Korean energy system during the 1960s and 1970s.

---

[1]   The Korean term denotes family-owned conglomerates that oversee a network of different companies. Often, individual *chaebol* groups oversee a wide variety of operations from agriculture to commercial, financial and manufacturing production, resembling in this respect the Western corporate conglomerate and the Japanese *zaibatsu*. *Chaebols* in South Korea are usually owned by a family and operate on the basis of strong familial loyalty and mutual trust that effectively coordinates the diverse activities of member companies and hence concentrates economic power. Hyundai, Samsung, Daewoo, Sunkyong, Goldstar and Ssangyong are representative *chaebols* which rank among the largest companies in the world.

Born of a centralist colonial heritage and assembled under national government direction, a South Korean "energy regime" (Winner, 1982) has been established in which the state, the public utility, conglomerates, and large foreign corporations have combined to create a centralized network of energy and industrial production. In addition to KEPCO, the South Korean energy system is dominated by a few *chaebols* who interact closely with the government and foreign companies. Several *chaebols* were able to build engineering and heavy industry companies into their conglomerate structures as a result of their power generation involvements. These conglomerates in turn became the principal sources of work for medium and small scale firms, which over time, have become hierarchically attached to the *chaebols* as suppliers of power plant equipment. Since the 1970s, the state, KEPCO and South Korean conglomerates have been able to set in place a firm technological base for the continued centralization of the energy-related sectors of the national economy.

Through the combined efforts of these private and public institutions, Japan and South Korea achieved rapid technicization in the energy sector. Initially dependent upon foreign capital and technology, the countries pursued a strategy of technology transfer with regard to power plant design, construction, engineering and management. Ambitious national plans called for a fully integrated electricity system, including rapid increases in installed generation capacity, the scale of power generation, the unit size of power turbines, and the efficiency of power generation and distribution, as well as the expansion of the transmission and distribution network. Most importantly, as unit sizes of power plants increased and facilities were clustered at aggregate sites detached from the load centers, electric choices were effectively isolated from the public arena.

In both countries, the alliance of the state, utilities and private conglomerates essentially removed economic and political decisions concerning the electricity system from effective community influence. This system of centralized power has made local communities dependent on these mega-organizations and their decisions, preempting community planning of the energy system, and impelling them to share the cost of the power network regardless of local effective benefits and costs (Messing et al, 1979). In this sense, the electricity system functions not only as a technological but political influence on both societies.

South Korea as a Case Study of a Nuclear Regime

While the evolution of a centralized energy sector prepared South Korean society for extensive technological development, it was the arrival of nuclear technology that fully integrated the nation's institutional, ideological and political structures by providing the practical setting in which a technocratic order could be designed and realized.   The formation of the nuclear technocracy and network was based on three factors: the forging of an institutional alliance among the state, the military, the *chaebol* sector and domestic science; the spread of an ideology of economic necessity and national security; and the establishment of a client relation between South Korea and the U.S.

KAERI and KEPCO established the Survey Committee on Nuclear Power Generation in 1962 at which time a "Plan for the Promotion of Nuclear Power Generation" was drawn up. The Plan recommended and rationalized nuclear power as the most promising power source to meet the "urgent" energy needs of "rapid" economic development in the near future. The plan cited the limits of anthracite coal, the major energy source at the time, for future power needs (KEPCO, 1965: 334).   Based on the committee's preliminary studies, South Korea established the Council on the Nuclear Power Generation Plan in 1965, with members from KEPCO, KAERI, the Ministry of Commerce and Industry, the Ministry of Construction, the Dahan Coal Corporation, the Korea Oil Company, and representatives from universities (Kim, 1967: 30). After conducting a comprehensive study of energy supply technology, the Council recommended the construction of two 500 MW nuclear power plants. In their report, the Council stressed the urgent need to add new electricity capacity and cited nuclear's projected decreasing costs and improving reactor safety. Responsibilities for the construction and operation of the plants were assigned to three government agencies: KEPCO for the engineering, construction and operation of the plants; KAERI for the research and development of nuclear power technology, fuel and safety controls, and the Economic Planning Board (EPB) for coordination of the South Korean nuclear power programme, negotiation of foreign loans and the preparation of feasibility studies (Ha, 1982: 225-226).

With the planning of the first power plant, Kori-1,[2] South Korea aggressively pursued the development of this technology with enthusiastic American support. While the U.S. denied South Korea access to atomic bomb technology, no such restriction was placed on nuclear power generation. After cancellation of a South Korea-French deal on a reprocessing pilot plant, KEPCO signed with Westinghouse and Atomic Energy of Canada, Ltd. (AECL) for two more power plants (Kori-2 and Wolsung-1) in 1977. For South Korea's fourth and fifth nuclear plants (Kori-3 and Kori-4), Westinghouse, the Export-Import Bank and the U.S. government collaborated to arrange the sales to South Korea in the increasingly competitive global nuclear market.

In the *chaebol* sector, the expertise and technology required for the construction and operations of nuclear power plants was being continuously accumulated. Nuclear power projects gave rise to a *chaebol* hierarchy, whereby attached small- and medium-companies could easily be transformed to nuclear equipment manufacture in the late 1970s. In most respects, nuclear

---

[2]   South Korea contracted with Westinghouse Electric International Company (WEICO) in 1970 as a prime contractor for the turnkey-based construction of the first nuclear power plant.

power plants were treated as simply larger and somewhat more complex versions of the already indigenized fossil-fuel plants.

Thus, between 1957 and 1975, a domestic nuclear technocracy was formed with the assistance of foreign, especially U.S., reactor companies, the IAEA and the U.S. government. These technocracies also played essential roles in the establishment of a "nuclear network" consisting of university departments, engineering and heavy industries, and research centers. At the same time, engineering and equipment manufacturing technology was being developed by domestic power plant companies.

Within this institutional framework, nuclear technology was considered the most promising technology not only for powering the country's future economic development, but also for securing the nation's defense. Propelled by a desire of the people to escape the misery of poverty and war experienced throughout the colonial period and the Korean War, economic and technological development was embraced as the dominant social priority over other social values. The military dictatorship frequently sought to justify its own actions as the necessary price of economic and technological progress.

The development of sophisticated nuclear power technology was expected to promote economic growth and security. It was also believed that a direct relationship existed between prosperity and energy consumption, with economic growth requiring ever-increasing amounts of energy. Many in the government believed that nuclear power was the only energy source capable of providing an "abundant energy machine" (Byrne and Rich, 1986) to meet the energy needs of rapid industrialization. Particularly since the two oil shocks of the 1970s, South Korea's energy security has been treated as synonymous with nuclear power because of the country's previous dependence on imported oil for electricity generation.

White smoke billowing from a reactor was therefore widely recognized in South Korea as a symbol of scientific progress. In concert with other developing nations, South Korea was committed to nuclear ideals and the indigenization of nuclear power plants. As Poneman points out (1982: 123):

> Large projects often appeal to developing country governments as a means to demonstrate their ability. Because of its complexity, perhaps, even its mystery, the mastery of nuclear technology can instill popular pride as well as enhance the legitimacy of a central government.

Nuclear power signaled the transition from under-development to development in the minds of South Korean leaders. The national government expected nuclear power to help in the consolidation and extension of its authority throughout the industrial economy in the same manner that rural electrification has forced the farming populace to rely on state leadership for an essential service. In this respect, nuclear technicism was seen as a complement to the ideologies of economism and political centralism that were practiced by the leadership.

The development of nuclear weapons was likewise regarded as a progressive decision, enhancing the security of the country and protecting it from possible invasion by the North. The perception of constant threats from North Korea was used to rationalize policies which elevated values of national security over civil society and accelerated the militarization of the country. Military-oriented technical and industrial development was emphasized and nuclear technology offered a perfect fit promising both the achievement of economic development and the means for military security and autonomy.

From the time that South Korea joined the IAEA in 1957, the government had been interested in developing nuclear weapons technology. Two events — the fall of South Vietnam in 1975 and President Carter's proposal in 1977 to withdraw American troops from the peninsula — are commonly speculated to have spurred South Korea to begin a "Manhattan Project" of its own as both a deterrent against invasion by the North, and as a bargaining chip in negotiations over the withdrawal of U.S. forces from South Korea.

According to Ha, aside from the problem of securing fissionable material, South Korea had little difficulty in mobilizing enough engineers and experts as well as equipment, and to make atomic bombs (Ha, 1978: 1139, 1141). But, because of the lack of sophisticated fuel-cycle technology, and the high cost associated with the production and fabrication of enriched uranium, the government sought to produce plutonium by construction of a reprocessing plant in partnership with France. The project was canceled in 1976 when the U.S. threatened to withhold export licenses and credits necessary to acquire American reactors, and finally threatened to terminate U.S. military supplies (Ha, 1982: 237).

Despite this decision, research and development on nuclear fuel-cycle technology remained a top priority project of the military government through the 1970s, which recognized that the successful implementation of its ambitious nuclear power program and the future possibility of weapon development depended on a stable domestic supply of nuclear fuel. The Korea Nuclear Fuel Development Institute (KNFDI), which was established in 1976 and later merged with KAERI, promoted the development of the nuclear fuel-cycle technology. The KNFDI completed a fuel-fabrication pilot plant and developed pilot plants for uranium refining and conversion with financial and technical help from France. Another approach to obtain fissionable material was the construction of a Canadian CANDU reactor. A reactor of this design was opened for operation in 1983. The low fissile waste from other light water reactors can be recycled in CANDU reactors, thereby making reprocessing and plutonium separation unnecessary. Furthermore, CANDU reactors are themselves well suited to the production of weapon-grade plutonium, presenting a greater possibility of domestic production of nuclear arms (Duderstadt and Kikuchi, 1979: 144).

After several experiences with U.S. intervention in its nuclear policy, South Korea attempted to diversify the source of its nuclear reactors and equipment as well as the supply of its nuclear fuel. Thus, the South continued its interest in Framatome as a source of technical assistance and fuel supply (including the development of its seventh and eighth plants), but was faced again with U.S. pressure. After president Carter's visit to South Korea in 1979, the country selected Westinghouse as the prime contractor for the seventh and eighth plants. The country was forced to accept its status as a client state of the U.S. and bow to restrictions on nuclear development, placing nuclear policy even further beyond the influence of South Korean civil society.

It was in this political and economic climate that increasing demand for self-sufficiency in civilian nuclear power technology arose. The government encouraged *chaebols* like Hyundai and Daewoo Engineering to become involved in the country's nuclear projects. In addition, Korea Power Engineering (KOPEC), a KEPCO subsidiary, was established in 1976 to specialize in nuclear engineering services. With the establishment of this technocratic and institutional network, the nuclear indigenization process was fully engaged. The creation of KOPEC completed the institutional alliance between a public utility (KEPCO), the state (MOST and MOER), science centers (KAERI) and selected conglomerates such as Hyundai and Daewoo.

Within this alliance, KEPCO was given sole responsibility for the construction and operation of all nuclear power plants. The agency used its monopoly status to forge collaboration agreements with numerous foreign vendors for such things as the sale of design documents, the participation of domestic firms in the production of specific equipment, and foreign technical assistance or training of South Korean engineers and workers. Central services such as project management, design and construction engineering were identified with specific foreign companies by KEPCO, and then on-the-job training overseas was arranged. Foreign engineering companies such as Westinghouse, Bechtel and Framatome agreed to provide not only the basic technical documentation and information but also offered personnel training at home facilities.

For example, the collaboration agreements with Westinghouse for nuclear power projects Nos. 10 and 11 called for training of KOPEC engineers in project management, engineering design and the operation of safety systems. KEPCO designated KOPEC as the prime architect-engineering contractor for these works which in effect served to promote and concentrate domestic capacity in nuclear power engineering. KOPEC conducted feasibility studies, engaged in nuclear power planning and engineering, and provided basic training of technical personnel for all units from 1976. Principally through the mechanism of collaboration agreements, KOPEC has sought to lower learning costs and shorten learning times. While South Korea still depends on foreign contractors and consultancy firms in the areas of design, safety and system engineering, these areas too are slated to be indigenized.

Domestic scientific and technical support to KOPEC has been continuously provided by KAERI. This institute undertook research and development on fuel technology, reactor technology and safety aspects of nuclear power plants. KAERI's Nuclear Safety Center (NCS) provides support to the government in the area of regulation and licensing and prepares safety analysis reports from inspections on nuclear power plants. NSC is also charged with setting nuclear safety standards. In addition to KAERI, the Korea Institute of Science and Technology (KIST), and four university departments of nuclear engineering, the Defense

Development Agency and the Korea Nuclear Development Corporation support the nuclearization effort.

Chaebols, particularly Hyundai, Daewoo and Samsung, established heavy industrial companies from their engineering or construction companies in order to participate in the nuclear campaign. But the national government decided in 1978 that the government-owned Korean Heavy Industry Company (KHIC) (which was formed by government buyout of Hyundai International and Daewoo Heavy Industries) would have an exclusive monopoly on nuclear construction. KHIC monopolized all nuclear engineering and equipment production, forcing *chaebols* to move to the international nuclear market. More recently, numerous domestic firms have been pressuring the government to reopen the domestic nuclear market while simultaneously accelerating their efforts to enter the export market and diversify into other energy technology fields.[3]

Challenges to Nuclear Power in East Asia

The evolution of highly centralized and technicized power complexes, built up over many decades, has prepared East Asia for an extensive nuclear program involving power reactors, research reactors and nuclear fuel cycle facilities. Centralized arrangements for nuclear power plant construction and equipment manufacture have been rationalized as necessary for the reduction of national dependence on imported oil, and as a basis for continuous economic and technological development and the preservation of each country's competitive edge in the global economy. Technological and economic centralism, aided by state planning,

---

[3] KHIC has a long-term training collaboration agreement with Combustion Engineering (USA) for boiler design technology, and seeks similar arrangements with foreign manufacturers of turbines and generators. Hyundai Heavy Industrial Co. is also accelerating the buildup of design capacity through expanded overseas training programs for its design engineers.

now seems to be so pervasive that it represents a powerful force in contemporary energy and development policy in Japan, South Korea, Taiwan and China.

Generally left out of this success story is a critical examination of the social reconstructions required by the technology. As South Korea's experience demonstrates, the "hard path" energy option (Lovins, 1977) depends on large-scale generation systems financed by huge outlays of capital which can only be provided by government or corporate conglomerates. Only a few experts can govern the technology, and society is exposed to unimaginable risks. These phenomena can be expected to cause mutual distrust and public alienation, opposition and even war. As Lovins points out, the list of demands imposed upon society is impressive (1977:148):

> [T]he hard path . . . demands strongly intervention-ist central control, bypasses traditional market mechanisms, concentrates political and economic power, encourages urbanization, persistently distorts political structures and social priorities, increases bureaucratization and alienation, compromises professional ethics, is probably inimical to greater distributional equity within and among nations, inequitably divorces costs from benefits, enhances vulnerability and the paramilitarization of civil life, introduces major economic and social risks, reinforces current trends toward centrifugal politics and the decline of federalism, and nurtures — even requires — elitist technocracy whose exercise erodes the legitimacy of democratic government.

There is little doubt that hard-path advocates have achieved significant success in East Asia. However, growing regional and domestic problems are now evident which call into question the long-term viability of the technology. Among the most serious problems are nuclear waste management, the use of plutonium-based reactor systems by Japan, and regional tensions caused by the ambiguous state of the North Korean nuclear program and the increasing reliance by South Korea on nuclear power.

Plant Siting and Waste Disposal

South Korea and Japan are confronting serious problems in finding sites for new reactors as well as for storing the waste that each country's nuclear power system is currently generating. Both governments are facing strong, and even violent, protest against proposed new generating and waste disposal sites. Indeed, the handling and disposal of the entire spectrum of nuclear waste represents a major stumbling block for Asia's growing nuclear powers. The strength of the resistance movement has led South Korea to try a number of tactics designed to gain public acceptance for both new nuclear plants and proposed waste sites. The measures, including sponsoring visits of legislators and local leaders to operating low-level waste (LLW) disposal facilities in Japan, Britain, France and Sweden, have met with marginal success. Thus, an attempt to locate a spent fuel interim storage site on Ahn Myun Do, a small residential island in the southern Yellow Sea of Korea, was halted in the wake of strong opposition by local residents. Similar efforts in other parts of South Korea have also encountered strong local resistance.

To cope with these challenges, the central government is attempting to coordinate the actions of the nuclear regime, including KEPCO, KAERI, the Korea Atomic Industrial Forum (KAIF), research institutes and even supportive scholars, in a public relations campaign designed to increase support for nuclear power expansion. It appears likely, however, that resistance will be stronger than before as the social demands for political decentralization and local authority grow. The South Korean transition to more democratic rule has also been accompanied by a generally critical analysis of past military governments and their policies, including their commitments to nuclear power. As a result of public resistance to the siting of new facilities, of the six new reactors currently under construction (13 are scheduled for construction by 2006), four are being built at existing sites; the remaining proposed new rectors are currently without host sites. In addition, spent fuel and radioactive waste from operating plants is being maintained on site. Most of these sites will soon reach their maximum capacity.

Japan faces many problems, including, especially, storage siting. Unlike South Korea, however, Japan has chosen to solve the problem, in large part, through an aggressive reprocessing program and the use of reprocessed plutonium in fast breeder reactors (FBR). According to Kato, the requirements of these systems are substantial (1993: 45):

> To smooth shifts to FBRs, plutonium handling technology and experience must be established prior to FBR commercialization . . . We must acquire "reprocessing" technology and experience to take plutonium out and as well as those of "fabrication" to manufacture plutonium fuels.

The development of these systems has led to significant regional unease, with many countries worrying about the use of fissile material for military weapons. One recent report, for instance, claims that Japan has sufficient capabilities to produce hundreds of nuclear weapons (Han, 1995: 46-49). Japan has gas centrifugation and atomic vapor laser isotope separation enrichment techniques that amount to the most advanced state of technology in this field. According to Han's analysis, these advanced reprocessing and enrichment capabilities will enable Japan to have materials that are convertible to military uses. Other concerns center around potential security problems (i.e., theft by terrorist organizations) and the environmental hazards posed by plutonium-based reactors, including the risk posed by the shipment of plutonium from France to Japan. While Japanese leaders have indicated that they are prepared to make marginal concessions in answer to critics, such as tightening up supply-demand accounting and providing transparency, it is clear that they will not tolerate attempts by foreign governments — especially the U.S. — to interfere with Japan's plutonium use program (*Nucleonics Week*, 1994:23).

The North Korean Problem

North Korea began to develop nuclear weapons in the early 1960s. By the late 1980s, however, a number of factors, including concerns about rising costs of trying to match more advanced South Korean conventional weapons capabilities, cutbacks in Soviet arms supplies, the distancing of the Soviet leadership from its old North Korea ally and movement towards South Korea, and the failure of the Chinese to fill the void left by the Soviet departure, combined to lend new urgency to the nuclear program (IISS, 1991-1992:138). Efforts were primarily focused on the development of military applications, including the development of delivery systems such as SCUD B and C missiles, as well as intermediate range ballistic missiles such as the Rhodong No. 1.

Within the last few years, the International Atomic Energy Agency (IAEA) has discovered major discrepancies between Pyongyang's stated and actual nuclear programs. Subsequent to these findings, international pressure has been applied to North Korea to abandon its nuclear program. Aware of the fact that there is no effective way to avoid the request for special inspection by the IAEA, North Korea announced its withdrawal from the NPT. After long and difficult negotiations, North Korea, Japan, the United States and South Korea have now agreed upon a program whereby North Korea will receive two reactors, to be built by South Korea and financed by Japan, South Korea and the United States.

The fear of North Korean nuclear weapons provides South Korea with a strong incentive to develop its own nuclear weapons program. Whether or not the South already possesses nuclear weapons capability, an assumption routinely made by many analysts in the region, there are now strong public voices urging a re-examination of the policy against the development of nuclear reprocessing and uranium enrichment capabilities, as well as the country's membership in the NPT. Japan has also expressed serious concern about the potential for nuclear proliferation throughout the region. According to Han, for instance, as long as the Chinese nuclear weapons program remains uncontrolled, Japan may be tempted to maintain the capacity to transform its civilian nuclear technology into a weapons program (Han, 1995).

The region's problems in siting new generating stations, its inability to cope with increasing amounts of both low- and high-level waste, and the ambiguous political intentions of the region's primary political and nuclear powers, are adding to the regional unease over the spread of nuclear technology. Further, the level of unease is likely to increase significantly in the coming years, as countries grapple with group security arrangements and the declining influence of the United States in the region.

Conclusion

Many in the West view East Asia as a model of successful industrial development. Left out of this image is the means used to achieve rapid growth, including the momentum toward centralized technicization. The development of nuclear power is perhaps most indicative of the region's reliance on centralist and technocratic institutions to direct development. Industrialization at the pace that it has occurred in the area may not have been achievable without strong commitment to centralized technicization.

While it is true that the countries of East Asia exhibit a tremendously varied set of economic and social forms, all demonstrate roughly the same pattern of nuclear development. Each pursued a similar development path in creating a commercial nuclear power system. They steadily established centralized forms of energy and a nuclear regime either for commercial power or for military weapons. In doing so, all have also consistently expressed faith in the ability of large-scale, centralized technology to deliver social and economic prosperity. Framed by this ideological commitment to technology (see Byrne and Hoffman in this volume), the requirements of continuous economic growth and, in some cases, the need to reduce national dependence on imported oil, the continued expansion of nuclear power is likely to be a major political issue in the coming years. Unfortunately, these countries may find that far from solving basic problems, nuclear power may have only created a new and probably more intractable set of social conflicts.

References

Byrne, John, et al. 1992. "Energy and Environmental Sustainability in East and Southeast Asia." *IEEE Technology and Society*. Volume 10, No. 4(Winter):21-29.

Byrne, John and Daniel Rich. 1986. "In Search of the Abundant Energy Machine." In John Byrne and Daniel Rich (eds). *The Politics of Energy Research and Development*. Volume 3, Energy Policy Studies. New Brunswick, NJ: Transaction Press.

Duderstadt, James J. and Chihiro Kikuchi. 1979. *Nuclear Power: Technology on Trial*. Ann Arbor, MI: The University of Michigan Press.

Ha, Young-Sun. 1982. "Republic of (South) Korea." In J.E. Katz and O.S Marwah (eds). *Nuclear Power in Developing Countries*. Lexington, MA: Lexington Books.

_____. 1978. "Nuclearization of Small State and World Order: The Case of Korea." *Asian Survey (18 November):* 1134-1151.

Han, Yong-Sup. 1995. *Nuclear Disarmament and Non-Proliferation in East Asia.* New York, NY: United Nations Institute for Disarmament Research.

Hilgartner, Stephen, Richard C. Bell, and Rory O'Connor. 1983. *Nukespeak: The Selling of Nuclear Technology in America.* Harmondworth, Middlesex, England: Penguin Books.

IISS. 1991-1992. *Strategic Survey.*

Japan Electric Power Information Center. December, 1994. *Electric Power Industry in Japan: 1994/1995.*

Kang, Chang Su. 1993. "Prospect for Nuclear Cooperation in Northeast Asia." Paper presented at the International Symposium on Nuclear Power. Seoul, Korea. June 25.

Kato, Masski. 1993. "A Consideration of Nuclear Power Generation with the 21st Century in Sight." *Energy in Japan* (November). The Institute of Energy Economics, Japan: 28-45.

Kim, Duck-Sun. 1967. *The Present and Future of Nuclear Power Programme.* The Korea Electric Association (Chungi Huophoji). No. 7 (May).

Kim, Jong-dall and John Bryne. 1990. "Centralization, Technicization and Development on the Semi-periphery: A Study of South Korea's Commitment to Nuclear Power." *The Bulletin of Science Technology and Society.* Volume 10, No.4: 212-222.

Korea Energy Economics Institute. 1995. *Korean Energy Review Monthly.* (April).

Korea Electric Power Company (KEPCO). 1965. *Electric Power Annual.* Seoul: KEPCO.

Lovins, Amory B. 1977. *Soft Energy Paths: Toward a Durable Peace.* Cambridge, MA: Ballinger Publishing Company.

Messing, Marc, H. Paul Friesema and David Morell. 1979. *Centralized Power: The Politics of Scale in Electricity*

Generation. Cambridge, UK: Oeleschlager, Gunn and Hain.

Nucleonics Week. 1994. "Outlook on Asian Nuclear Power: Special Report to the Readers of Nucleonics Week." June 30 and July 4 issues. New York: McGraw-Hill.

Poneman, Daniel. 1982. *Nuclear Power in the Developing World.* London, UK: George Allen & Unwin.

Winner, Langdon. 1982. "Energy Regimes and the Ideology of Efficiency." In George H. Daniels and Mark H. Rose (eds). *Energy and Transport: Historical Perspectives on Policy Issues.* Beverly Hills, CA: Sage Publications.

_____. 1977. *Autonomous Technology: Technics Out-of-Control as a Theme in Political Thought.* Cambridge, MA: MIT Press.

Contributors

John Byrne is director of the Center for Energy and Environmental Policy and professor of energy and environmental policy at the University of Delaware. He has served as an advisor to the Korea Energy Economics Institute, the Finnish Department of Energy, the Chinese Academy of Sciences, and the Tata Energy Research Institute (India). He is co-editor of *The Politics of Energy R&D* and *Energy and Environment: The Policy Challenge*, and is the principal author of *Toward Sustainable Energy, Environment and Development*, a 1991 volume prepared for the World Bank.

Cate Gilles is a journalist and photographer focusing on Native American issues in the Four Corners region of the American Southwest. She was news director of the Dine Bureau for the Gallup Independent from 1994 to 1995..

Phillip A. Greenberg is an independent consultant on energy and environmental policy based in San Francisco, California. Previously, he served as Senior Staff to two subcommittees in the U.S. House of Representatives, where he focused on energy and nuclear power. Prior to his service in Washington, he was Assistant for Energy and Environment to California Governor Edmund G. Brown, Jr. He also served as chair of the California State Task Force on Nuclear Energy and Radioactive Materials and was a member of the Governor's Review Panel on Nuclear Power Plant Emergency Preparedness following the Three Mile Island accident. He authored the first comprehensive study of nuclear waste generated in California.

Michael T. Hatch is a professor of political science and international studies at the University of the Pacific, Stockton, California. He is the author of *Politics and Nuclear Power: Energy Policy in Western Europe*. His current research is on the interactions of domestic politics and international negotiations in recent efforts to address the global warming question.

Steven M. Hoffman is associate professor of political science at the University of St. Thomas, St. Paul, Minnesota, where he directs the University's Environmental Studies program. He is also an adjunct research professor at the Center for Energy and Environmental Policy of the University of Delaware. He is co-editor of *Energy and Sustainable World Development* (a special issue of *Energy Sources*) and has published numerous articles on energy and environmental policy. He is also active in environmental policy in Minnesota through his role in several state-wide advocacy and policy organizations.

James Jasper is associate professor of sociology at New York University. He is primarily interested in the moral, emotional, and cognitive dimensions of political action. His works include *Nuclear Politics* (Princeton University Press, 1990), *The Animal Rights Crusade* (co-author, The Free Press, 1992), and *The Art of Moral Protest* (University of Chicago Press, forthcoming).

Jong-dall Kim is assistant professor of economics at the Kyungbook National University, Taegu, South Korea. He is also an adjunct research professor at the Center for Energy and Environmental Policy of the University of Delaware. He previously served as a senior policy analyst for the Korean Energy Economics Institute.

David Marples is professor of history at the University of Alberta, Canada. He is the author of four books. His newest work, *Belarus: From Soviet Domination to Nuclear Catastrophe* (1996) is published by Macmillan Press (UK).

Cecilia Martinez is assistant professor of ethnic studies at Metropolitan State University, St. Paul, Minnesota. She is also an adjunct research professor at the Center for Energy and Environmental Policy of the University of Delaware. Her principal research interests include energy and environmental policy, and the political economy of social inequality, especially as it pertains to issues of environmental justice. She has worked with the U.S. Department of Housing and Urban Development and was a member of the Sustainable Communities Initiative sponsored by the Minnesota Environmental Quality Board.

Carolyn Raffensperger is a lawyer and has graduate degrees in anthropology and archeology. She worked for nine years for the Sierra Club on many of the major environmental issues of the 1980s. She served on the Illinois Low Level Radioactive Waste Disposal Facility Siting Commission from 1990 to 1993 and is presently a member of the National Academy of Sciences' Committee on Decommissioning and Decontamination of the Uranium Enrichment Facilities. At present, she directs the Science and Environmental Health Network.

Printed and bound by CPI Group (UK) Ltd, Croydon, CR0 4YY

23/10/2024

01777672-0005

Index

Advisory Committee on
Uranium 69
Ahn Myun Do 291
Argentina 31, 33
Argonne National Laboratory
83
Armenia 252, 26
Atomic Energy Act 80 ff
energy weapons and re-
search, 80; free-market
rationales, 86, 90; revision
of the Act, 87
Atomic Energy Commission
(AEC) 80-83
and Lewis Strauss, 11, 51;
enabling legislation, 80;
oversight of mining, 105;
and the Atomic Energy Act
of 1946, 135; and the
JCAE, 203-204
Atomic Energy of Canada, Ltd.
283
Atoms for Peace
Eisenhower's initiation of,
1, 84, 135; and the Man-
hattan Project, 83-85; and
South Korea, 279

Babcock and Wilcox 88

Belarus 8; and Chernobyl
evacuation 254, 256;
levels of morbidity 259;
future nuclear policy 262-
263
Below Regulatory Concern
(BRC) 190-191
Beloyarsk station (Russia) 250
Big Science 91
and the AEC laboratory
system, 91-97
birth defects 109, 116
Brazil 31, 33
Brookhaven National
Laboratory 83, 146
Bush, Vannevar 70-72, 96-97

Calvert Cliffs 142, 206
CANDU reactor 27, 40, 286
chaebol (South Korea) 280, 287
Chernobyl 2, 8, 11; and
continuing viability of
nuclear power 12; Soviet
response to, 26; Chinese
response to, 39; French
response to, 60; and
Churchrock spill, 108;
public trust, 128; protest
actions, 145; nuclear